AF757174

Elements of Reinforced Concrete Design

(with an introduction to prestressed concrete)

624.012.45 LAF

Olu Lafe, Department of Civil Engineering, University of Lagos

MACMILLAN
PUBLISHERS

© O. Lafe, 1986

All rights reserved. No reproduction, copy or transmission of this publication may be made without written permission. No paragraph of this publication may be reproduced, copied or transmitted save with written permission or in accordance with the provisions of the Copyright Act 1956 (as amended). Any person who does any unauthorised act in relation to this publication may be liable to criminal prosecution and civil claims for damages.

First published 1986

Published by *Macmillan Publishers Ltd*
London and Basingstoke
Associated companies and representatives in Accra, Auckland, Delhi, Dublin, Gaborone, Hamburg, Harare, Hong Kong, Kuala Lumpur, Lagos, Manzini, Melbourne, Mexico City, Nairobi, New York, Singapore, Tokyo

ISBN 0-333-41942-1

Printed in Hong Kong

624.012·45 LAF

Contents

Preface to the Series

The African Network of Scientific and Technological Institutions exists to bring about close and active collaboration between university-level institutions in training and research. The ANSTI administrative centre is at the UNESCO Regional Office for Science and Technology for Africa in Nairobi. It has established sub-networks in nine engineering fields and four basic science fields. Its objectives include

— strengthening communications through meetings, conferences, seminars, publications and study tours;
— promotion of training in technological research and postgraduate studies through senior staff exchanges and visits, exchange of students, fellowships, in-service training of teachers, and support for the acquisition of practical skills;
— promotion of research and development in areas of significance and relevance to the region.

As part of the first objective listed above, ANSTI publishes a biennial African journal of Science and Technology, and the ANSTI Newsletter. In addition, ANSTI encourages the writing of textbooks.

The provision of affordable and appropriate textbooks has always been a problem in the region, and it is a problem which has often been exacerbated by economic crises. The increasing variety of courses being taught in member institutions also increases the need for textbooks. ANSTI started a few years ago to encourage teaching staff at member institutions to write textbooks for their courses, and found funds to help subsidise the cost. The result was the publication, with Macmillan Publishers, of *Engineering Hydraulics* by Professor J.M.K. Dake and *Structural Analysis* by Professor N. Tebedge.

The success of these texts has been a source of encouragement to accelerate the publication of textbooks, each with its own distinctive design, in a series known as the ANSTI Technology Series. As before, the books in the series are subsidised low cost student editions. Several titles in civil engineering are under development and it is hoped to expand to other fields in future.

Professor E.B. Kwakye
ANSTI/RAIST
Nairobi
September, 1986

Preface

Through his own experience as a student and a teacher, the author has observed that for those beginning a structural design course the most trying aspect is perhaps that of understanding the basic concepts. Some students, despite a sound knowledge of the theory of structures, find the essential methodology of design extremely difficult to grasp. The impression they have is that the design process borders almost on a trial-and-error game with no well-defined, consistent procedure to be followed. Other students are so used to designing rectangular sections that a minor change in geometry to, say, a trapezoidal section (such as often occurs in some bridge structures) may create enormous problems.

The general aim of the book is to present a unified view of the processes involved in the design of reinforced (and prestressed) concrete structures. Considerable emphasis has been placed on the analysis of sections from first principles as this is an important component of the design process.

Design charts have been carefully avoided in this particular book for two major reasons. Firstly, the use of design charts, the author believes, is a rather poor way of teaching the elements of design. Secondly, modern design offices are increasingly computerising the design process and the development of the appropriate software programs is normally based on the basic design equations, not on the charts derived from them.

The text covers both the so-called working stress and limit state design methods. The requirements of the Nigerian Code of Practice (NCP 1, 1973, Part I — *The Structural Use of Concrete in Buildings*), are followed for the working stress (ie, the modular-ratio and the load-factor) methods. For the limit state design the British Code of Practice CP 110 has been followed.

Chapter 1 contains a broad overview of the major aims of reinforced concrete design, including important definitions, basic forms of structural deformations, descriptions of materials, plus an overall view of the existing methods of design. Chapters 2 to 6 cover the use of the modular-ratio or the elastic method for the design of reinforced concrete structures.

Readers interested essentially in the limit state design method are still encouraged to go through these initial chapters (at least cursorily) as some of the concepts introduced in them would be found useful in later study.

Chapter 7 covers the load-factor method and shows how important design formulae can be obtained when the basic assumptions are followed.

The limit state design method is presented in Chapters 8–11. The arrangement and general set-up of the material contained in these chapters follow closely those for the working stress methods, but obviously repetition (eg, the design of bases) has been avoided. Thus, the bulk of the content of Chapter 6, which deals with the design of foundations by the working stress methods, will also be applicable to design by the limit-state method, and no special chapter has been dedicated to the design of bases in part two of the book.

Chapter 12 provides an introduction to the design of prestressed concrete structures, with the pertinent design equations being, as much as possible, established from first principles. Chapter 13 covers the general precepts of structural detailing and shows simple samples of detail drawings for common structural elements.

In the last chapter, 14, the important topic of computer-aided design is discussed. Descriptions and listings of some design computer programs developed by the author and his students over the years are also presented.

The author is indebted to the following people who, one way or another, have helped during the preparation of the manuscript: Prof. N. Tebedge (University of Addis Ababa); Prof. C.O. Orangun; Prof. T.A.I. Akeju (University of Lagos); Mr. T. Badmus; Ms. O. Oremodu; Mr. H.K. Johnson; Mr. A. Onigbanjo; Mr. I. Uka; Mr. J. Oteri (former civil engineering students at the University of Lagos); Mr. A.A. Adefala who effected a peerless typing of the original manuscript; and last, but not the least, Ms. I. Lafe, the author's wife, who was a consistent source of encouragement during the write-up.

O.L.

Lagos, April 1986.

1 General introduction

1.1 Structures

A structure may be defined as any system that has the sole function of transmitting load. The structure may consist of a single element or a combination of several elements. Some engineers use the term 'structure' to denote a group of elements, and refer to the single elements as 'members' of the structure. In this book both usages have been freely adopted but the reader is not left in doubt as to what is implied in each particular case.

1.1.1 Classification of structures

A structure may be characterised by its *support system* (eg, a simply-supported structure, a cantilever, etc); its *shape* (eg, a shell structure, an arch, etc); or the behaviour of its *material* under load (eg, a linear structure, a non-linear structure, etc).

A **simply-supported structure** is one that has moment-free supports at its ends (Fig. 1.1). One common example of a simply-supported structure is a concrete beam resting on a block wall.

A **fixed-end** (or built-in) **structure** is one for which the moments at the ends are non-zero (Fig. 1.2). The fixed ends help to relieve the moments and reduce the deflection within the span. Fixity is often achieved in slabs cast *monolithically* with supporting reinforced concrete beams. Two concrete elements are said to be monolithic when, during construction, fresh concrete is poured at the same time for both members such that when the concrete is cured (or set) no joint separates one element from the other. Monolithic elements behave as a single element.

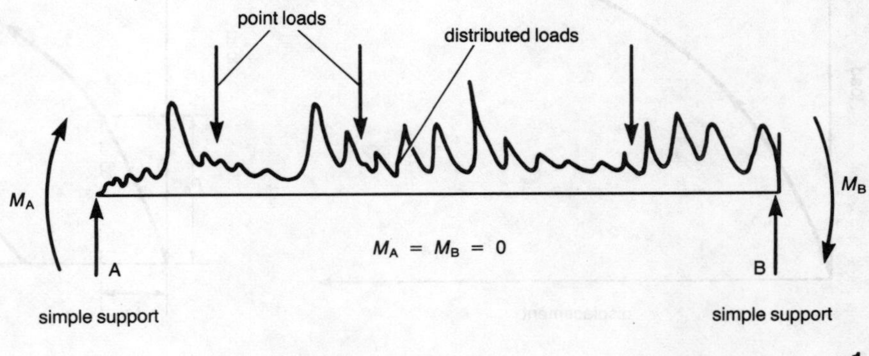

Fig. 1.1 Simply supported structure

Fig. 1.2 Fixed-end structure

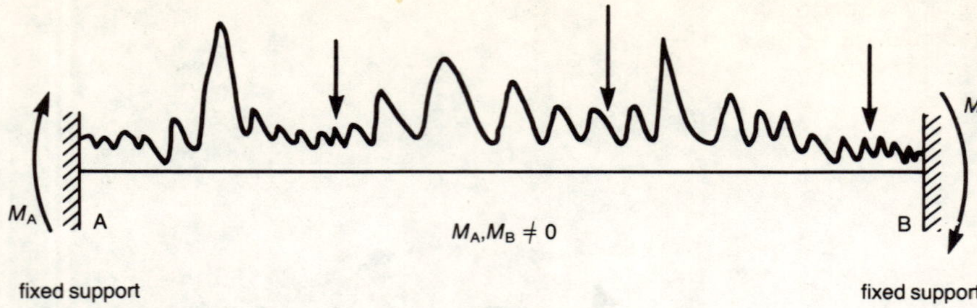

Fig. 1.3 *below*. Cantilevered structure

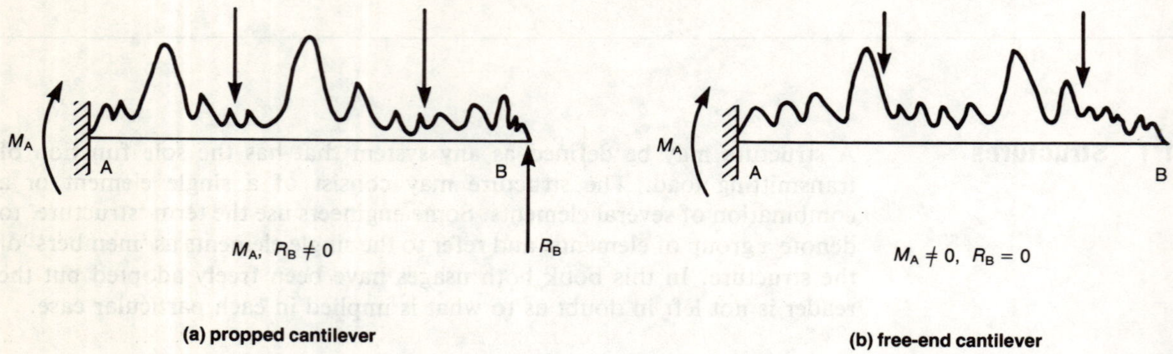

Fig. 1.4 *right*. Continuous structure

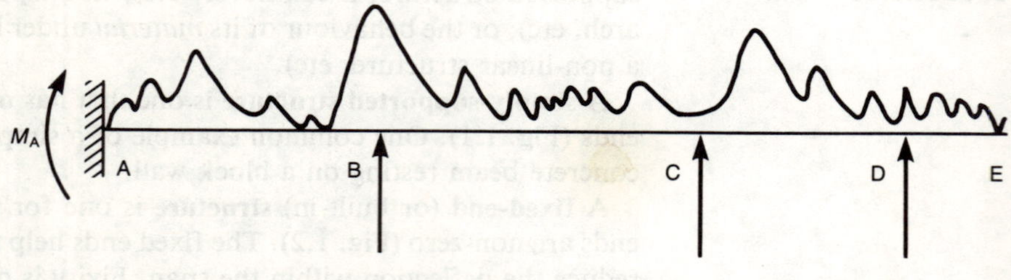

Fig. 1.5 *below left*. Load/displacement curve for an elastic structure

Fig. 1.6 *below right*. Load/displacement curve for an inelastic structure

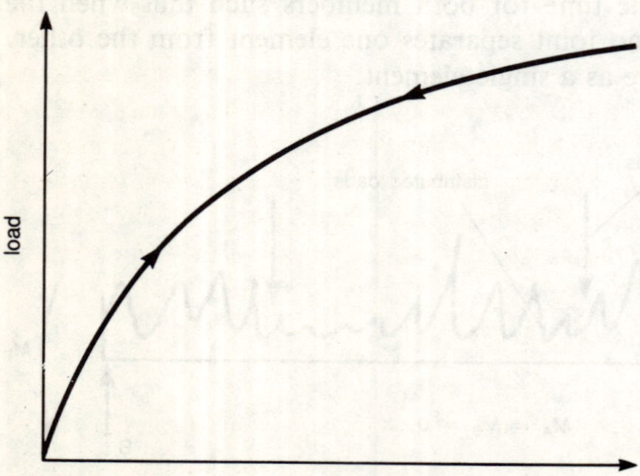

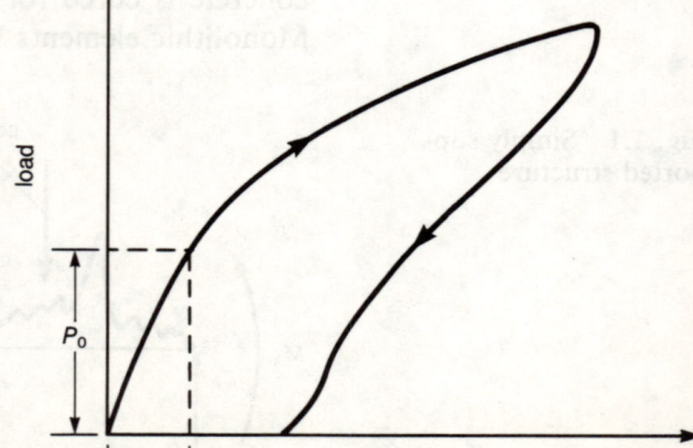

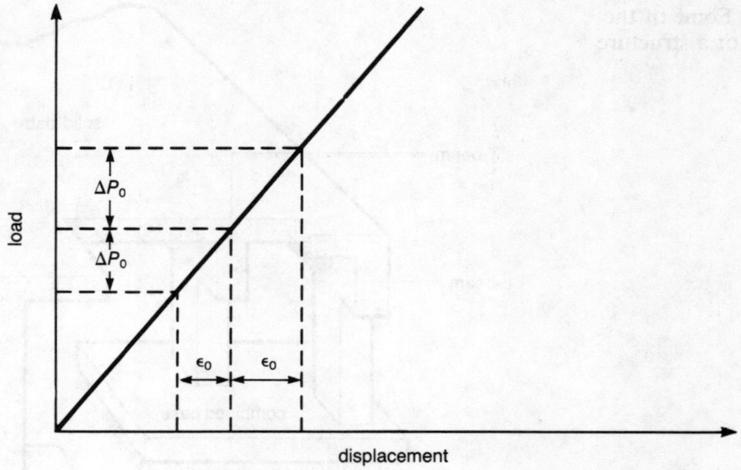

Fig. 1.7 Load/displacement curve for a linear structure

A **cantilever** is a structure which is fixed at one end but has either (a) a simple support or (b) no support at all at the other. The first type is called a propped cantilever and the other is appropriately termed a free-end cantilever (Fig. 1.3).

When a structure has more than two supports it is referred to as a **continuous structure**. The extreme ends of such structures may be free, fixed or simple (Fig. 1.4). Continuous structures occur frequently in practice. However, in order to simplify analysis, designers often break the continuous structure into the simpler, single-span elements described earlier.

An **elastic structure** is one which, when loaded, still has the ability to return to its original unloaded condition. The load/displacement graph for an elastic structure is of the form shown in Fig. 1.5.

On the other hand an **inelastic structure** is one that cannot be brought back to its original condition once loaded. A typical load/displacement curve for an inelastic structure is shown in Fig. 1.6.

Depending on the range of loading and/or deformation suffered by an inelastic structure it may often be possible to operate within the elastic region of the load/displacement curve. For example, the load P_0 giving rise to a displacement ϵ_0 in Fig. 1.6 will allow the structure to return to its original unloaded condition. This is the main justification for the assumption of elastic behaviour in some of the design methods described later.

A structure (or its material) is **linear** when it suffers equal deformation under correspondingly equal loading increments. The load/displacement graph is thus a straight line, as shown by the typical graph in Fig. 1.7. A non-linear structure experiences unequal deformation for every equal loading increment.

1.2 Structural elements

A good understanding of the various elements of a given structure is necessary in order to make full use of existing design codes and to appreciate those structural systems commonly referred to in this book.

Fig. 1.8 Some of the elements of a structure

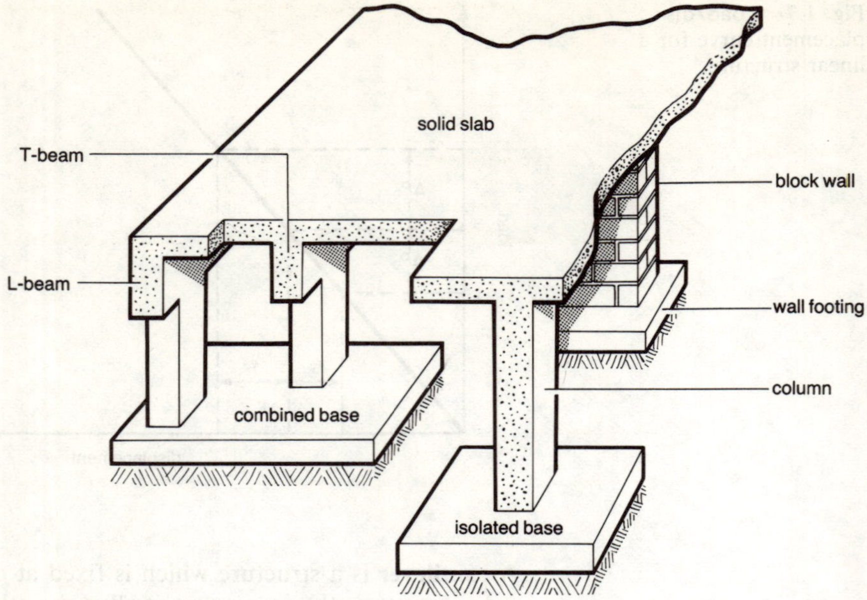

1.2.1 Slabs

A **slab** is a planar structural system having an areal extent that is comparatively larger than its overall thickness. A slab is usually constructed in the horizontal plane (Fig. 1.8); slabs which are placed vertically are better referred to as walls.

Theoretical analysts in the field of solid mechanics refer to some slabs as 'plates'. Some of the terms used to describe different types of slabs are: solid, ribbed and flat.

A **solid slab** is just that — solid (without voids), as indicated in Fig. 1.8. Examples of ribbed, hollow-block or voided slabs are shown in Fig. 1.9.

Ribbed, hollow-block and **voided slabs** have one major advantage (over solid slabs) in that electrical conduits, lighting components and plumbing pipes can be concealed more easily within the structure.

The term **flat slab** is commonly used for a reinforced concrete slab that is not generally supported by beams but rests directly on columns (Fig. 1.10). The slab may have drops and the column heads may be flared; such features help to reduce the punching shear suffered by the slab.

Flat slabs with drops and flared column heads are more suitable for huge industrial structures (eg, warehouses) and others for which the proportion of the imposed load far exceeds that of the dead load. On the other hand, flat slabs without drops and flared column heads are better for short-span structures with relatively light imposed loads. This second category of flat slabs (ie, those with constant thickness and without flared column heads) are commonly referred to as **flat plates**. Generally, flat plates are more appealing aesthetically than those with drops and flared column heads.

Flat slabs have several advantages over regular solid slabs. These include:

- better illumination, because there are no beams;
- greater freedom in locating columns, because these do not have to be in line to support beams;
- smaller dead load on the super-structure and the foundation;

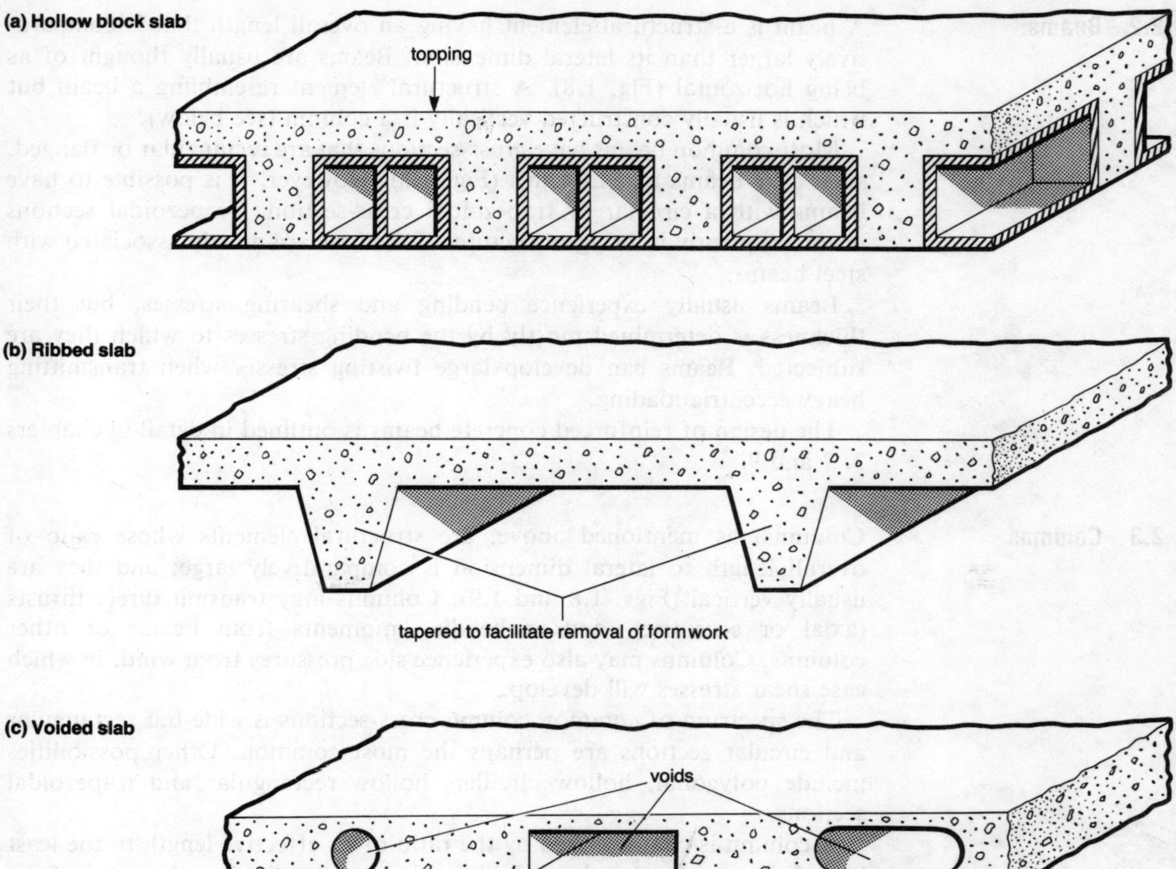

Fig. 1.9 *above*.
Examples of ribbed, hollow block and voided slabs

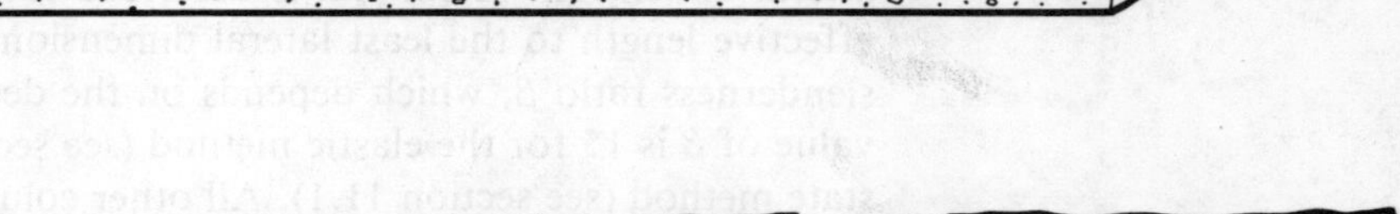

Fig. 1.10 Flat slab

- greater ease in installing utility items such as sprinklers, pipes, electrical conduits, etc;
- elimination of complicated formwork for beams.

Bending is usually the most important consideration when designing a slab but, for flat slabs, shearing at the column heads is usually the most critical factor. The analysis and design of reinforced concrete slabs is covered more comprehensively in chapters 4 and 10.

1.2.2 Beams

A **beam** is a structural element having an overall length that is comparatively larger than its lateral dimension. Beams are usually thought of as being horizontal (Fig. 1.8). A structural element resembling a beam but which is usually constructed vertically is a column (see below).

Most common beams have cross-sections that are rectangular or flanged, such as T-beams and L-beams (Fig. 1.8). However, it is possible to have beams with a circular or trapezoidal cross-section. Trapezoidal sections occur frequently in bridge structures. I-sections are mostly associated with steel beams.

Beams usually experience bending and shearing stresses, but their thickness is determined mostly by the bending stresses to which they are subjected. Beams can develop large twisting stresses when transmitting heavy eccentric loading.

The design of reinforced concrete beams is outlined in detail in chapters 3, 7 and 9.

1.2.3 Columns

Columns, as mentioned above, are structural elements whose ratio of overall length to lateral dimension is comparatively large, and they are usually vertical (Figs. 1.8 and 1.9). Columns may transmit direct thrusts (axial or eccentric), and/or bending moments from beams or other columns. Columns may also experience side pressures from wind, in which case shear stresses will develop.

The spectrum of common column cross-sections is wide but rectangular and circular sections are perhaps the most common. Other possibilities include polygonal, hollow circular, hollow rectangular and trapezoidal sections.

A column is characterised by the ratio of its effective length to the least lateral dimension. A column is described as 'short' when the ratio of the effective length to the least lateral dimension is less than a certain critical slenderness ratio β, which depends on the design method being used. The value of β is 15 for the elastic method (see section 5.1) and 12 for the limit state method (see section 11.1). All other columns whose slenderness ratios exceed β are said to be 'long' or 'slender'.

Further information on the analysis and design of reinforced concrete columns is presented in chapters 5 and 11.

1.2.4 Foundations

The **foundation** of a given structure is a structural system (a base) by which the load from the structure (often termed the super-structure) is transmitted to the soil in the most efficient way. Whenever the load intensity that would be imposed on the ground by a given structure would exceed the pressure that the soil could safely bear, a foundation must be provided. The foundation helps either to redistribute the load more uniformly over the ground or to serve as the medium through which the load is transferred to a firmer soil below ground level.

The type of foundation provided for a particular super-structure would depend on the type of structure, the magnitudes and areal distribution of the load, the nature of the soil and, invariably, on economic considerations. Common foundation types include the isolated base, the combined base, the raft foundation and the piled foundation. Isolated and combined bases are shown in Fig. 1.8.

An **isolated base** is a footing for an individual column, whereas a **combined base** supports a group of two or more columns. Combined bases are usually provided either when it is impracticable to support individual columns on isolated bases or when there is reason to believe that doing otherwise would cause some of the columns to settle more than the rest (ie, differential settlement may occur).

A **raft foundation**, as the name suggests, is essentially a footing in the form of a raft. Raft foundations are commonly used where ground pressures are so low that the provision of isolated or combined bases would be unsafe and/or uneconomical. A rule of thumb is that once the total area of all bases required to support the structure exceeds 50 per cent of the available ground surface then it would be expedient to combine the footings into a single 'giant' footing or a raft.

A raft may be in the form of a solid element [Fig. 1.11(a)], a beam-and-slab raft [Fig. 1.11(b)], or a cellular raft [Fig. 11(c)]. The observant reader should have noticed a striking similarity between the classification of normal slabs (section 1.2.1) and that of raft foundations.

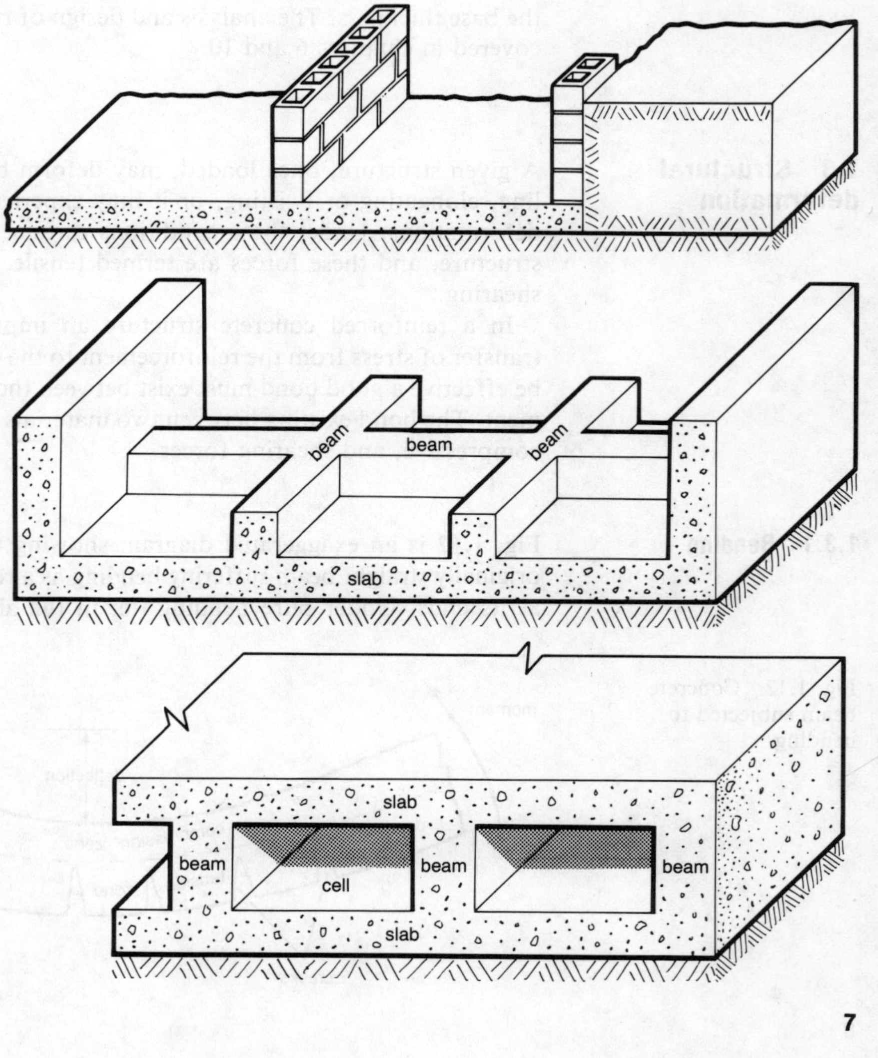

Fig. 1.11 Solid, beam-and-slab and cellular rafts

(a) Solid raft

(b) Beam-and-slab raft

(c) Cellular raft

A solid raft is used when the thickness of base required (with full consideration of the bending and shear effects) to support the imposed loads is not excessive (300 mm is generally regarded as the upper limit). When a greater thickness is required (eg, when punching shear requirements are considered) a beam-and-slab raft could be more appropriate. The cellular raft is usually provided when the total depth required is high (more than 1000 mm), and has the advantage that the cells can be used for storage, electrical conduits, plumbing works, etc.

When the soil is incapable of carrying the load imposed by the structure, a means is sought by which the load can be transmitted to a firmer or more resistant soil or rock below ground level **(point-bearing piles)** or be carried by the frictional forces between the soil and the pile **(friction piles)**. Sometimes a group of piles (a **pile group**) is provided to supplement a raft foundation system.

More comprehensive information on the behaviour and design of piles can be found in most good textbooks on soil mechanics (see References, for example: Terzaghi and Peck, 1967).

Bending and shearing are the most critical factors in the design of foundations, with shearing usually being the critical condition for determining the base thickness. The analysis and design of reinforced concrete bases are covered in chapters 6 and 10.

1.3 Structural deformation

A given structure, once loaded, may deform by bending, shearing, buckling, elongating or twisting, or it may experience a combination of these deformations. Under these conditions certain forces are developed in the structure, and these forces are termed tensile, compressive, torsional and shearing.

In a reinforced concrete structure an important consideration is the transfer of stress from the reinforcement to the concrete. For the transfer to be effective a good bond must exist between the concrete and the reinforcement. The bond existing between two materials is determined by the tensile, compressive, and shearing forces.

1.3.1 Bending

Fig. 1.12 is an exaggerated diagram showing the typical behaviour of an originally straight beam suffering bending as a result of the moments applied at its ends. Under pure bending (ie, in the absence of axial thrusts) the

Fig. 1.12 Concrete beam subjected to bending

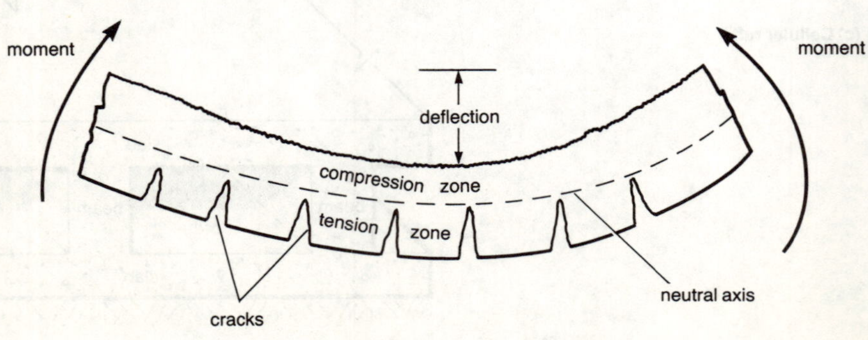

Fig. 1.13 Beam subjected to direct shear

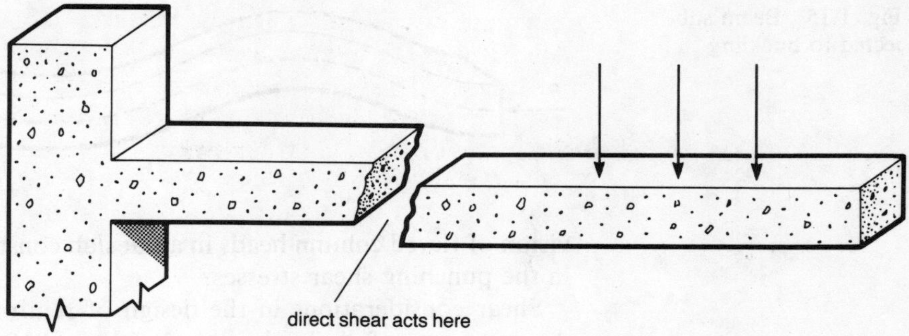

Fig. 1.14 Base subjected to punching shear

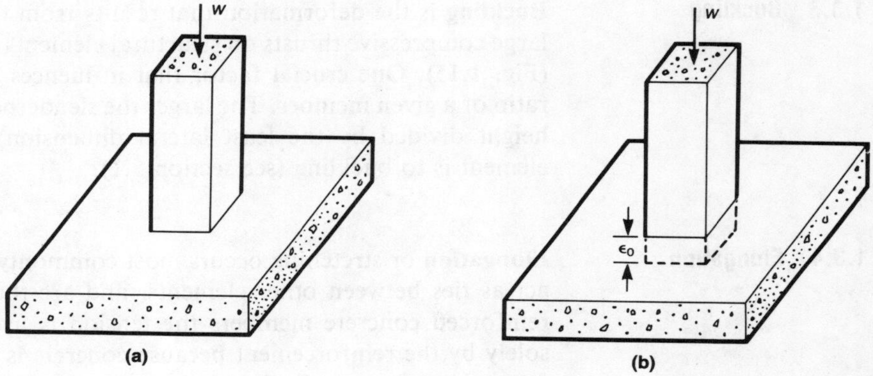

upper portion of the beam will be in compression while the lower part is in tension. The plane on which there exists neither tension nor compression is called the **neutral axis.**

In concrete beams (and slabs) cracks will usually develop in the tension zone. The deflection produced by the bending, if excessive, can have undesirable effects on finishes and may even alarm the users. The deflection limits acceptable for the design of beams and slabs are presented in sections 3.2, 4.9, 9.2 and 10.5.

1.3.2 Shearing

In any structure that is transmitting lateral loads, certain forces having a tendency to separate or "shear" off some sections of the structure from the rest develop. The magnitude of the shear forces generally varies from one part of the structure to another.

There are two major types of shear — **direct shear** and **punching shear**. When a beam carries vertical loads shearing forces are developed across the sections and these tend to break the beam into two parts, as depicted in Fig. 1.13. Under such conditions the beam is experiencing direct shear.

Whenever a concentric load acts on a slab or a footing (usually through a column) there is a tendency for the load to 'break through' or punch the slab or footing (Fig. 1.14). The magnitude of the punching shear stress varies directly with the magnitude of the transmitted load but inversely with the perimeter of the punching surface. This is the main reason why the pro-

Fig. 1.15 Beam subjected to buckling

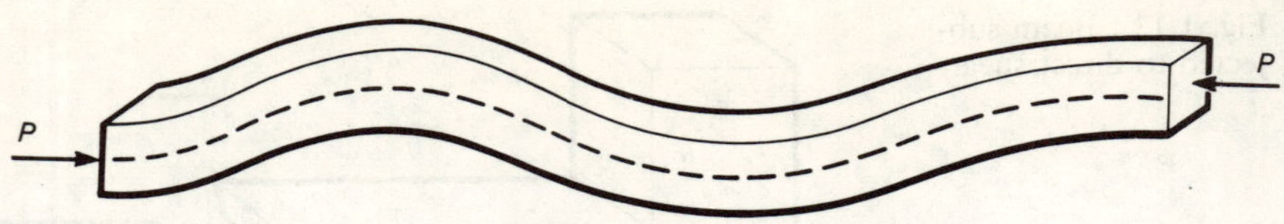

vision of flared column heads in a flat slab construction leads to a reduction in the punching shear stresses.

Shear considerations in the design of reinforced concrete members are discussed more fully in chapters 3, 6, 9 and 10.

1.3.3 Buckling

Buckling is the deformation that results from the application of relatively large compressive thrusts on structural elements such as beams and columns (Fig. 1.15). One crucial factor that influences buckling is the slenderness ratio of a given member. The larger the slenderness ratio (effective length or height divided by the least lateral dimension) the more susceptible the element is to buckling (see section 5.1).

1.3.4 Elongation

Elongation or stretching occurs most commonly in structural elements that act as ties between other elements, and are thus subjected to tension. In reinforced concrete members the tension is usually assumed to be taken solely by the reinforcement because concrete is weak in tension (hence the formation of cracks in the tension zone shown in Fig. 1.12).

1.3.5 Twisting

Fig. 1.16 shows the twisting effect produced on a beam by an eccentric load transmitted by an adjoining cantilever. A concrete member must be adequately stiffened (by the provision of sufficient main and link reinforcement) in order to be able resist twisting forces.

Fig. 1.16 Beam subjected to twisting

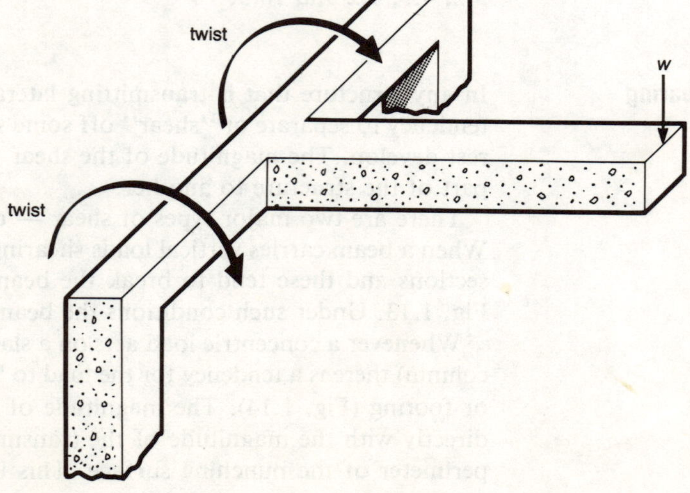

1.4 Concrete

Concrete is a mixture of cement, fine aggregate (eg, sand), coarse aggregate (eg, gravel), and water. Plain concrete, which has an average unit weight of 22.0–23.6 kN/m^3 (depending on the type of aggregate used) is brittle and weak in tension. Although there exists no constant relationship between the tensile and compressive strengths of concrete, the former is about one-tenth to one-thirtieth of the latter.

The tensile strength of concrete is commonly neglected in the design of most ordinary structural elements. However, in the design of some structures that are required to contain liquids the tensile strength is taken into account.

The strength of concrete is determined partly by the relative proportions of the cement, and of the fine and coarse aggregates, but the water/cement ratio, which is discussed in section 1.4.4, is another important strength factor.

The compressive strength of concrete is determined by performing compression tests on standard sizes of concrete blocks or cylinders. In some countries (eg, Nigeria, Great Britain) 150 mm cubes are used, while in others (eg, USA, some European countries) cylinders with a diameter of 150 mm and a height of 300 mm are the standard sizes used.

The common practice is to make two sets of three standard blocks or cylinders and to test one set after 7 days and the other after 28 days. Fig. 1.17 shows the test results obtained by Orangun (1976) of the cube strength as a function of time for some concrete mixes in Nigeria. The results show clearly the importance of the brand of cement on the strength, as the mix proportions and water/cement ratios were the same for all cubes.

Fig. 1.17 Strength/age relationships for concrete mixes with different brands of cement. (Orangun, 1976)

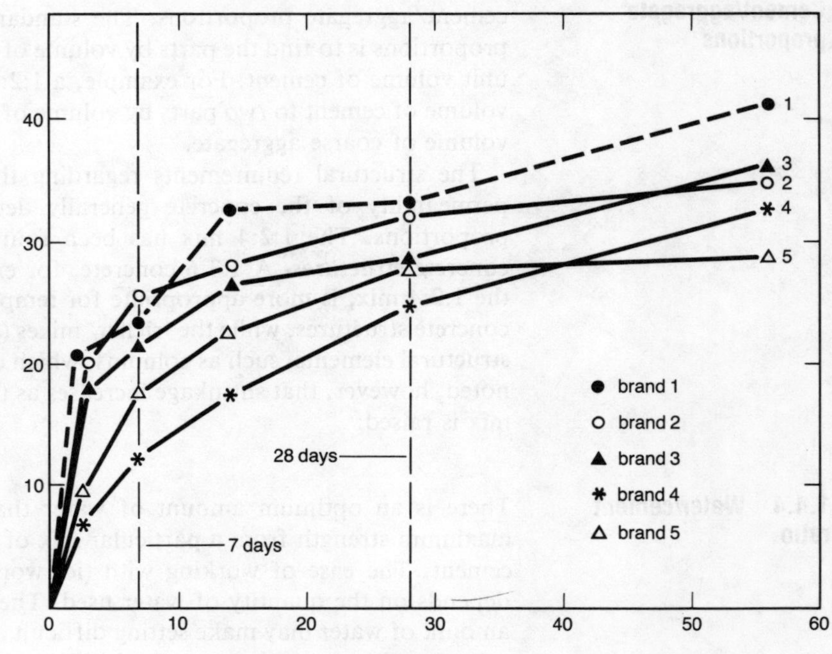

1.4.1 Cement

Various types of cement are available, depending on the constituents and process of manufacture. The most common is ordinary Portland cement which is manufactured by burning specific proportions of various compounds of calcium to form clinker. The clinker is then crushed into the powder-like form that is known as cement.

Ordinary Portland cement is used for general construction work where no specific properties are required. Other types of cement that are available for special applications include rapid-hardening Portland cement (which allows faster removal of formwork), sulphate-resistant cement (suitable for structures exposed to acidic soil and saline water) and low-heat Portland cement (which is less likely to shrink or crack and is therefore good for heavy structures such as dams).

1.4.2 Aggregates

Sand is generally the fine aggregate used for concrete, while coarse aggregate is usually either gravel or crushed stone.

Considerable research interest has been shown recently in the possible use of laterite as a fine aggregate. Preliminary results indicate that a relatively smaller amount of cement is required (for the same strength of concrete) if laterite is used as the aggregate. In addition, the insulating property of the mix is greatly improved. This is clearly an area in which research has definitely not caught up fast enough with practice, because stabilised laterite has been used for building structures in Nigeria for several years now.

The basic requirements for aggregate sizes, grading, and quality are specified in the appropriate design codes.

1.4.3 Cement/aggregate proportions

As stated earlier the strength of a particular concrete depends partly on the cement/aggregate proportions. The standard method of measuring these proportions is to find the parts by volume of fine and coarse aggregates to a unit volume of cement. For example, a 1:2:4 concrete implies *one* part by volume of cement to *two* parts by volume of fine aggregate to *four* parts by volume of coarse aggregate.

The structural requirements regarding the strength, life-span and impermeability of the concrete generally determine the cement/aggregate proportions. The 1:2:4 mix has been found suitable for most ordinary concrete structures. A 1:3:6 concrete, for example, which is 'leaner' than the 1:2:4 mix, is more appropriate for temporary (soon-to-be-demolished) concrete structures, while the 'richer' mixes (eg, 1:1:2) are more suitable for structural elements, such as columns, which carry large thrusts. It should be noted, however, that shrinkage increases as the proportion of cement in the mix is raised.

1.4.4 Water/cement ratio

There is an optimum amount of water that will produce a concrete of maximum strength from a particular mix of fine and coarse aggregate and cement. The ease of working with (ie, workability of) the concrete also depends on the quantity of water used. The use of less than the optimum amount of water may make setting difficult and reduce workability. On the other hand, greater shrinkage and a reduction in strength will occur when more water than the optimum amount is used.

The best water/cement ratio therefore depends on the particular concrete mix. Values running between 55 and 60 per cent have been suggested for the 1:2:4 mix commonly used for concrete structures. For best results a laboratory test called a **slump test** must be carried out on various mixes. The mix satisfying the prescribed maximum slump (about 150 mm for reinforced concrete) is selected.

1.5 Reinforcement

Concrete is commonly reinforced with steel bars which can be either plain or twisted, but in general any material that adequately satisfies requirements with regard to yield stress, ductility, tensile strength and other related and important properties can be used to reinforce concrete. To this end radical research work (notably in Nigeria and Japan) has recently been carried out to investigate the use of common cane and bamboo fibres as reinforcing materials.

There are various types of steel reinforcement used in practice, depending on the specific job requirements. Plain round mild steel bars are perhaps the most common and these generally have a minimum yield stress of 250 N/mm^2. Deformed mild steel bars have better bonding properties than plain ones, however. Other types include cold-worked bars (minimum yield stress 425 N/mm^2) and fabric reinforcement bars (these come in square and oblong meshes) which are commonly used for slabs (especially ground floor slabs) and certain shell structures.

The diameter of most common steel reinforcement varies between 6 mm and 40 mm. However, for most ordinary reinforced concrete structures the popular sizes used are those between 8 mm and 25 mm in diameter.

1.6 Design requirements

A good structural design must satisfy three basic requirements — those of strength, serviceability and economy. The requirement for strength is to ensure that the structure is capable of carrying the applied loads.

The requirement for serviceability relates to the amount of deflection, cracking and vibration that may be suffered acceptably by the structure and its users. The condition of the completed structure must be such that users are not alarmed by its appearance, and that the finishes are not damaged.

The final aim of the designer would be to maximise strength at minimum cost, within the bounds of specified serviceability conditions.

Among the several factors influencing the economic aspect of a structure are the cost of materials, the type of structure, and the conditions at the site (soil type, wind loads, ect). For reinforced concrete structures, the designer must obtain a judicial balance between the section size and the amount of reinforcement by means of a careful analysis of current market prices.

1.7 Balanced design

In respect of the optimal use of materials in a reinforced concrete structure, the best proportion is that at which, if there was to be a failure, both concrete and steel would fail simultaneously. In other words, if this optimal proportion is used the maximum resistance moments due to concrete and steel will be equal. In such a case the proportion of reinforcement is known as the 'economical' amount and the design is described as being 'balanced'. It is worth nothing that the term 'economical' used here has nothing to do with monetary costs as such.

When a reinforced concrete member has more reinforcement than the balanced or economical amount, the member is said to be 'over-reinforced'. A member is 'under-reinforced' when it has less than the balanced amount of reinforcement.

Over-reinforcement should be avoided as much as possible because the failure of such members is much more sudden than that of correspondingly under-reinforced members.

1.8 Codes of Practice

Most countries have standards that must be complied with by every structural designer in order to meet the strength, serviceability and other important local requirements. These standards are published in the form of Codes of Practice (commonly designated CP). For example, the BS CP 114 is a British Code of Practice for the design of reinforced concrete buildings in the United Kingdom. The NCP 1, published by the Nigerian Standards Organization (Federal Ministry of Industries), is the Nigerian equivalent of the BS CP 114.

The British Code of Practice, CP 110 (which replaced the older codes CP 114, CP 115 and CP 116) governs the structural use of concrete. This book is based broadly on the requirements of either NCP 1 or CP 110, but the particular standard used is clearly stated where necessary.

The newest British Code of Practice is BS 8110 which will be used to complement CP 110. It has only just become available in Great Britain.

1.9 Design methods

There are three major methods that are commonly used for the design of reinforced concrete structures. These are the **elastic** (or **modular-ratio**), the **load-factor** and the **limit state** methods of design. The first two are based on NCP 1 (or CP 114), and the third is based on CP 110 (the Nigerian equivalent of which has not been published yet).

1.9.1 The elastic method

The elastic method is the oldest of all the three design methods, and its main advantage is that it has been able to withstand the test of time. Experience has shown that, generally, structures are safe when designed according to the requirements of the elastic method.

In this method use is made of certain 'permissible' (or 'allowable') stresses in conjunction with an elastic analysis of the section. It is also assumed that the ratio *m* between the modulus of elasticity for steel and the corresponding value for concrete is constant. The permissible stresses are some fraction of the specified (or characteristic) strength of the concrete or steel.

The permissible stresses perform a variety of functions, such as giving an idea of the carrying capacity of the structure and providing indirect control of the serviceability criteria.

Failure is assumed to have occurred when the permissible stresses have been exceeded. The permissible stresses are not a measure of the maximum or ultimate load the structural member is capable of carrying.

There are some inherent assumptions that are made in the elastic method. These are:

- that plane sections remain plane after bending. This implies that strain varies linearly across the section in bending. (In fact this particular assumption is not limited to the elastic method because linear strain variations are also assumed in the other two design methods.);
- that stress is proportional to strain for both concrete and steel (that is, a constant modular ratio);
- that there is a perfect bond between the concrete and the steel;
- that temperature and shrinkage stresses are negligible. Such stresses should be taken into account, however, in irregular and long structures. In arches and membrane structures temperature often has a significant effect on the stresses. However, in most common reinforced concrete structures temperature stresses can be ignored quite safely because the thermal coefficients of expansion of concrete and mild steel are approximately equal (about 10×10^{-6} per °C);
- that concrete has negligible tensile strength. It is assumed that the concrete in the tension zone will develop cracks up to the neutral axis. (As stated earlier in section 1.4 this assumption must not be made for some structures designed to contain liquids; the tensile strength of the concrete must be taken into account in this case.);

The elastic method is treated in greater detail in chapters 2 to 6.

1.9.2 The load-factor method

A major shortcoming of the elastic method is the failure to represent adequately the behaviour of a member when failure is approached. The linear stress-strain relationship assumed becomes unrealistic when loads much higher than the working or service loads are being carried. The load-factor method (also called the ultimate-load method) was formulated to correct this obvious anomaly in the elastic (modular-ratio) method.

The load factor is essentially the ratio of the ultimate strength of a structural member to its working load. The basic precepts are:

- the overall load factor in respect of material strengths is 1.8;
- a larger factor of safety is used for concrete. This is to take into account the possibility of more variations in the strength of concrete (because the strength depends on several factors, eg, mix proportions, water/cement ratio, etc) than in the strength of steel and the important requirement that, in the case of failure, the steel should yield first, rather than have the concrete fail by explosive compression. To this end two-thirds of the actual cube strength is used for designed mixes and three-fifths is used for nominal mixes. This implies additional safety factors of 1.5 for designed mixes and 1.67 for nominal mixes;
- the stresses in the steel must not exceed the 'permissible stresses' (see section 2.2);
- a uniform compressive stress equal to two-thirds of the permissible compressive stress in concrete is used;
- for beams and slabs the depth of concrete in compression must not exceed half the effective depth (which is the distance between the far side

of the compression zone and the location of the tension reinforcement);
- the limiting strain in concrete at failure is 0.0033.

A linear stress-strain relationship is not assumed in the load-factor method and the modular ratio is not used. Further discussion of the load-factor method is presented in chapter 7.

1.9.3 The limit state method

The elastic and the load-factor methods of design both provide indirect controls for the serviceability requirements of a structure. With the limit state design method the strength requirements are separated from the serviceability requirements.

To satisfy strength requirements the ultimate load capacity of the structure is assumed to be much greater than the working loads. Two partial safety factors are used, one for material strength and the other for loads and load effects (CP 110, part 1: 1972). The serviceability requirements stipulate that:

- the deflection of the structure, or any part of the structure, should not adversely affect the appearance or efficiency of the structure;
- any cracking of the concrete should not adversely affect the appearance or durability of the structure;
- any vibration must be at a level that does not discomfort or alarm the users and must not cause damage to the structure or impair its efficiency.

The inherent assumptions in the limit-state method include the following:

- concrete has a negligible tensile strength (as in the modular-ratio method);
- plane sections remain plane after bending, or a linear strain distribution exists across the section (as in the modular-ratio method);
- special curves are provided (see section 8.2) for the stress-strain relationship in the concrete and reinforcement;
- the limiting strain in the concrete is 0.0035;
- for singly-reinforced sections (ie, only tension reinforcement provided) the depth of concrete in compression must not exceed half the effective depth.

The method recognises three types of load: the characteristic dead load G_K (ie, the weight of the structure along with finishes, partitions and fixtures); the characteristic imposed (or live) load Q_K; and the characteristic wind load W_K. Appropriate partial factors of safety are introduced for each load type, depending on the limit state being considered. These safety factors are supposed to account for:

- possible unusual increases in load beyond those considered in deriving the characteristic load;
- inaccurate assessment of the effects of loading, and unforeseen stress redistribution within the structure;
- variations in dimensional accuracy that occur during construction;
- the importance of the limit state being considered.

To obtain the design strength of a structure for a given material and a given limit state, the characteristic strength (eg, the cube strength of the concrete or the yield stress of the steel) is divided by the appropriate partial safety factor for strength (see section 8.3). The partial safety factor accounts for possible differences between the strength of the material in the actual structure and the strength derived from test specimens.

The limit state method is the newest of all the three design methods and it is considered to:

- present a more realistic view of the behaviour of the structure under varied conditions, especially when failure is approached;
- take a more direct approach in the control of the deflection, cracking, vibration, etc, of a structure;
- recognise the relative importance of load types on the various limit states;
- take into account the possibility of different material strengths, loading and other conditions to those used in the design;
- allow for the review of the partial safety factors for special structural problems (only the minimum safety and serviceability requirements have been suggested) and for the inclusion of additional limit states considered necessary by the engineer.

A more detailed treatment of the limit state method of design is given in chapters 8 to 11.

1.10 Structural analysis

A good structural design must be preceded by an equally good structural analysis. Structural analysis here is taken to mean the process by which the loads acting on a structure, the support system, and the resulting forces and moments are determined.

For any given rigid body that is in a state of static equilibrum the following conditions must be satisfied:

- the algebraic sum of all vertical forces must be zero;
- the algebraic sum of all horizontal forces must be zero;
- the algebraic sum of moments on the structure must be zero.

These conditions are known as the equations of a rigid body that is in static equilibrium, and any structure that can be analysed by these three questions alone is said to be **statically determinate**. Otherwise the structure is **statically indeterminate**.

Some examples of how to analyse particular structural systems are given in this book, whenever it is believed that they enhance the design process. However, as this is basically a textbook on design no attempt has been made to discuss the various aspects of structural analysis too deeply.

1.11 Exercises

1.1 A beam 5 m long rests on brick walls at its extreme ends. Is the beam: (a) an encastre member; (b) a built-in member; (c) a propped cantilever; (d) a simply-supported member?

1.2 The load/displacement curve for the material of a structure can be represented by the equation $L = 5.79d^2$ where L = load in kN and d = displacement in mm. Is the structure:
(a) linear; (b) linear but inelastic; (c) non-linear but elastic; (d) non-linear and inelastic?

1.3 Is modular ratio:
(a) the ratio of Young's modulus for concrete to that for steel; (b) the ratio of the compressive stress in concrete to the tensile stress in steel; (c) the ratio of Young's modulus for steel to that for concrete; (d) none of

the above?

1.4 In the elastic method of design is the modular ratio assumed to be: (a) constant; (b) slowly varying; (c) dependent on temperature stress; (d) equal to unity?

1.5 Is the neutral axis defined as a plane where: (a) the moment is zero; (b) the direct force is zero; (c) the stress is zero; (d) the direct shear is zero?

1.6 Is permissible stress: (a) the maximum stress a material is capable of developing; (b) a measure of the cube strength of concrete; (c) the yield stress of steel; (d) none of the above?

1.7 Is any structure that satisfies the strength requirement one that: (a) can carry any applied loads; (b) can never crack; (c) can never bend; (d) can transmit the design loads?

1.8 In the elastic method, it is assumed plane sections remain plane after bending. Does this mean:
(a) strain is always proportional to stress; (b) strain is constant over the section; (c) stress is constant across the section; (d) strain varies linearly across the section?

1.9 For most reinforced concrete structures temperature stresses are often ignored: Is this because:
(a) concrete is highly resistant to temperature changes; (b) concrete is weak in tension; (c) the percentage of steel in the section is too small; (d) concrete and steel expand/contract by about the same amount for the same temperature change?

1.10 In the modular-ratio method, is it assumed that concrete in the tension zone has failed because:
(a) it makes analysis easier; (b) the neutral axis must be located; (c) concrete cannot carry tension at all; (d) it is safe to do so although concrete can carry some tension?

1.11 If U = ultimate strength of a member and W = the working load is the load factor then:
(a) W/U; (b) U/W; (c) $(U-W)/(U+W)$; (d) none of the above?

1.12 In the load factor method of design is it true that:
(a) permissible stresses are unnecessary; (b) only the permissible stress in concrete is used; (c) only the permissible stress in steel is used; (d) permissible stresses in steel and concrete are used?

1.13 As far as the load factor method is concerned, if A = actual strength of concrete and C = cube strength of concrete, then is:
(a) $A = C$; (b) $A > C$; (c) $A < C$; (d) $A \leq C$?

1.14 When a structure satisfies the serviceability criterion, is it true that it can never:
(a) be engulfed in fire; (b) suffer any deflection; (c) sink or crack; (d) none of the above?

1.15 Does the limit state method:
(a) separate strength and serviceability requirements; (b) meet serviceability requirements by satisfying strength requirements; (c) use the load factor to satisfy serviceability requirements; (d) satisfy strength but not serviceability requirements?

1.16 Does the term 1:3:6 concrete mean:
(a) 1 part of cement to 3 parts of sand to 6 parts of gravel; (b) 1 part of

sand to 3 parts of gravel to 6 parts of cement; (c) 1 part of cement to 3 parts of aggregate to 6 parts of water; (d) none of the above?

1.17 Is it true that a continuous structure:
(a) is always indeterminate; (b) can be determinate; (c) is always determinate; (d) can never be solved by the equations of rigid body alone?

1.18 In the load factor method, are modular ratios:
(a) necessary; (b) unnecessary; (c) useful but dispensable; (d) assumed constant?

1.19 Is bamboo used in certain traditional buildings because:
(a) it is strong in tension; (b) it is strong in compression; (c) it is approximately circular in its natural form; (d) the bond stresses are large?

1.20 If a structure collapses, was:
(a) the original design not according to the Code of Practice; (b) the serviceability criterion exceeded; (c) the construction defective; (d) none of the above?

1.21 Must a free-end cantilever concrete beam be reinforced:
(a) below the neutral axis; (b) above the neutral axis; (c) on the neutral axis; (d) none of the above?

1.22 As far as most methods of design are concerned, is it assumed that concrete:
(a) can carry no moment; (b) can carry some tension in certain situations; (c) can carry compression only; (d) can carry tension only?

1.23 If P = permissible stress and S = the specified strength of a material, then is:
(a) $P \geq S$; (b) $P = S$; (c) $P \leq S$; (d) $PS = 1$?

1.24 When the elastic method is used to design a reinforced concrete structure, then is it true that:
(a) cracks will always form in the tension zone; (b) cracks may not necessarily form in the tension zone; (c) cracks are developed everywhere; (d) none of the above?

1.25 In the load factor method, if A = stress, B = strain and C = constant then does:
(a) $A = CB$; (b) $A = C/B$; (c) $A = C^2B$; (d) none of the above?

1.26 For the ultimate limit state, if F = partial factor of safety, then is:
(a) $F = 1$; (b) $F \leq 1$; (c) $F \geq 1$; (d) F = limit (stress/strain).

1.27 If A = the amount of cement in a 1:2:4 concrete and B = the amount of cement in an equivalent volume of 1:1:2 concrete, then is:
(a) A > B; (b) A < B; (c) A = B; (d) the relationship impossible to obtain?

1.28 Is it true that a reinforced concrete member with a balanced amount of reinforcement:
(a) cannot collapse; (b) cannot deflect; (c) is always cheap to construct; (d) none of the above?

1.29 Is a propped cantilever:
(a) determinate; (b) indeterminate; (c) somehow determinate depending on your point of view; (d) none of the above?

1.30 Using the elastic method of design, is it true that once the strength requirement is satisfied the structure:
(a) will not deflect at all; (b) will never collapse; (c) will not crack; (d) none of the above.

For answers to these exercises, please see p 221.

(a) $\gamma = 1$; (b) $\gamma \leq 1$; (c) $\gamma \geq 1$; (d) γ = limit (stress/strain)?

(a) $A > B$; (b) $A < B$; (c) $A = B$; (d) the relationship impossible to obtain?

Part One: **The Elastic Method and the Load-factor Method**

(designs based on the requirements of NCP 1)

2 Introduction to the Modular-ratio Method

2.1 Notation

The notation shown in Table 2.1 overleaf has been adopted largely to conform with the current generally-accepted international standards. Special notation is used where it is believed that to do otherwise would be ambiguous. For example, the symbol σ is generally used to represent the stresses in concrete or steel (with an appropriate suffix for each) when there is the need to emphasise that these stresses are not necessarily equal to the allowable stresses.

The symbol m is used throughout to denote the modular ratio, while the symbols x and z refer to the neutral-axis depth and the lever arm, respectively.

2.2 Assumptions

As stated earlier, in chapter 1, with the elastic (or modular-ratio) method plane sections are assumed to remain plane after flexure. This leads to a linear strain distribution across the section, with the obvious implication that strain varies directly with distance from the neutral axis.

As both the concrete and the steel are assumed to be elastic the stress-strain relationship is also linear within the limits of the permissible (or allowable) stresses. Because the tensile strength of concrete is ignored, all tensile stresses are assumed to be taken by the steel reinforcement. Special cases in which the tensile strength of concrete is taken into account include the design of liquid-containing structures and when the concrete is assumed to resist diagonal tension within the limits of shear stress specified. These must be understood before a meaningful analysis/design of structural members can be achieved.

2.3 Permissible stresses

The elastic method is based on the use of certain permissible stresses. These permissible stresses not only serve as measures for the strength of the structural member but they also determine (albeit indirectly) the serviceability requirements.

Table 2.1 Notation for the modular-ratio method

Symbol	Quantity
A_{st}	area of tension reinforcement
A_{sc}	area of compression reinforcement
a	lever-arm factor
b	width of section
b_w	rib or web width of a section
d	effective depth to tension reinforcement
d'	effective depth to compression reinforcement
E_c	modulus of elasticity of concrete
E_{st}	modulus of elasticity of steel
e	eccentricity
F_c	total compression carried by concrete (or flange only)
f_c	compression carried by rib of a flanged beam
F_{sc}	force carried by compression reinforcement
F_{st}	tension carried by tension reinforcement
h	overall depth of section
h_f	thickness of flange
K_c	moment of resistance factor based on concrete
K_{st}	moment of resistance factor based on steel
l_x	length of shorter side of a rectangular slab
l_y	length of longer side of a rectangular slab
M	maximum applied moment
M_R	moment of resistance of section
m	modular ratio ($= E_{st}/E_c$)
n	neutral axis factor ($= x/d$)
p_c	permissible stress in concrete (in bending)
p_{cc}	permissible stress in concrete (indirect compression)
p_{cu}	cube strength of concrete
p_{sc}	permissible stress in compression reinforcement
p_{st}	permissible stress in tension reinforcement
p_{sy}	yield stress in steel
w	uniformly-distributed load per unit area
x	depth to the neutral axis
z	lever arm
ϵ_c	maximum strain in concrete
ϵ_{sc}	strain in tension reinforcement
ρ	proportion of tension reinforcement ($= A_{st}/bd$)
ρ'	proportion of compression reinforcement ($= A_{sc}/bd$)
ϕ	bar diameter
σ_c	maximum compressive stress in concrete
σ_{sc}	stress in compression reinforcement
σ_{st}	stress in tension reinforcement

Table 4, NCP 1 provides the permissible stresses for nominal concrete mixes and these are quoted here in Table 2.2.

The permissible tensile stresses in mild steel reinforcement should not exceed 126 N/mm^2 (NCP 1: clause 3.4.2). However, for steels with a guaranteed yield stress, the permissible stress can be increased to 50 per cent of the guaranteed yield stress but it must not exceed 210 N/mm^2.

The permissible compressive stresses in steel reinforcement should not exceed 112 N/mm^2 (NCP 1: clause 3.4.3). Again, with steels having a guaranteed yield stress, the permissible stress may be increased to 50 per cent of the guaranteed yield stress but it must not exceed 175 N/mm^2.

The Code does allow the permissible stresses in concrete and in the

Table 2.2 Permissible stresses for nominal concrete mixes

mix proportions	permissible concrete stress				
	compressive			bond	
	direct (N/mm²)	due to bending (N/mm²)	shear (N/mm²)	average (N/mm²)	local (N/mm²)
1:1:2	7.6	10.0	0.9	1.0	1.5
1:1.5:3	6.5	8.5	0.8	0.93	1.40
1:2:4	5.3	7.0	0.7	0.83	1.25

reinforcement to exceed those given above, provided the increase is limited to 25 per cent and provided that:

- the excess is due solely to wind-induced stresses;
- the stress in the reinforcement never exceeds 250 N/mm^2.

3 Beams

3.1 Members subjected to bending

Beams may be deformed either by bending or by shearing action. In the following section the attention of the reader is focussed on beams that are subjected to bending moments. The requirements for designing beams to resist shear are outlined in section 3.3.

3.1.1 Singly-reinforced sections

Members that are reinforced for tension only are said to be singly-reinforced. Normally, reinforcement for concrete structures is provided to help take the stresses in the tension zone. However, as will be shown in section 3.1.2, the concrete may also be reinforced in the compression zone under certain special conditions.

Now, consider a typical rectangular section [Fig. 3.1(a)] of a member subjected to bending. The linear stress distribution over the section is shown in Fig. 3.1(b). Note that no curve is drawn for a large portion of the tension

Fig. 3.1 Singly-reinforced rectangular member

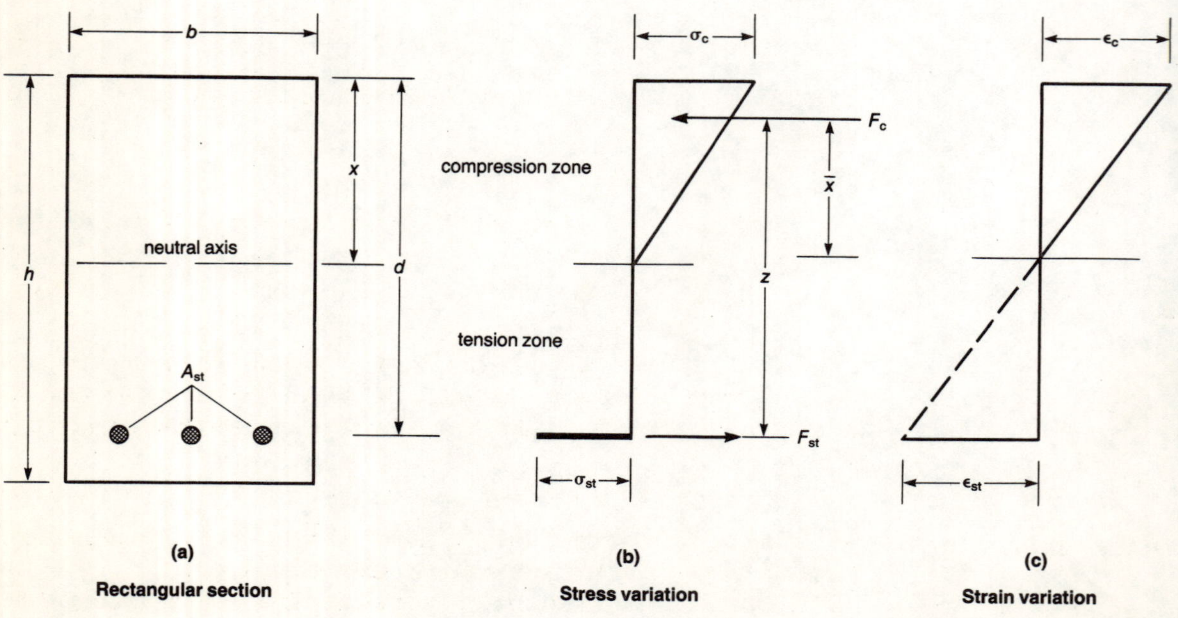

zone (except at the location of the reinforcement) because the concrete in this region is assumed to have failed. Fig. 3.1(c) shows the linear strain variation across the section.

The position of the neutral axis can be obtained by comparing the two similar triangular strain distributions in the tension and compression zones [Fig. 3.1(c)]. That is:

$$\epsilon_c/x = \epsilon_{st}/(d-x) \tag{3.1}$$

from which

$$x/d = n = 1/(1+\epsilon_{st}/\epsilon_c) \tag{3.2}$$

As $\epsilon_{st} = \sigma_{st}/E_{st}$
and $\epsilon_c = \sigma_c/E_c$
equation (3.2) can be written in the form:

$$n = 1/(1+\sigma_{st}/m\sigma_c) \tag{3.3}$$

where $m = E_{st}/E_c$ is the modular ratio.

Equation (3.3), which is independent of the shape of the section, is an important formula for design. For a 'balanced' design (see section 1.7), the stresses σ_{st} and σ_c will be equal to the permissible stresses p_{st} and p_c in steel and concrete, respectively, and equation (3.3) becomes:

$$n = 1/(1+p_{st}/mp_c) \tag{3.4}$$

For the equilibrum of the section, the total compression must be balanced by the total tension. That is:

$$F_c = F_{st} \tag{3.5}$$

The compression F_c is the integrated sum of all the elemental forces $\mathrm{d}F_c$ acting on an element (area $b\mathrm{d}y$) at a distance y from the neutral axis. As the stress on this element (by proportion) is equal to $(y/x)\sigma_c$, the total compression is given by:

$$F_c = {}_0\int^x (y/x)\sigma_c \times b\mathrm{d}y = \tfrac{1}{2}\sigma_c bx \tag{3.6}$$

By using equation (3.6) in equation (3.5), and setting $F_{st} = \sigma_{st}A_{st}$ we obtain:

$$\sigma_{st}/\sigma_c = bx/2A_{st} \tag{3.7}$$

By eliminating σ_{st}/σ_c in equation (3.3), via equation (3.7), and dividing through by bd, the result is:

$$n^2 + 2m\rho n - 2m\rho = 0 \tag{3.8}$$

which is a quadratic equation suitable for determining the neutral axis factor n, when the steel proportion ρ is known. Equation (3.8) can be solved for n to give (after rejecting the obviously unrealistic negative root):

$$n = \sqrt{m\rho(m\rho + 2)} - m\rho \tag{3.9}$$

which can be used whenever the position of the neutral axis is to be located, given the section size and proportion of reinforcement.

The moment of resistance M_R of the section is the internal couple the section is capable of transmitting. This couple is equal to $F_c z$ or $F_{st} z$. That is:

$$M_R = \tfrac{1}{2}\sigma_c bxz = A_{st}\,\sigma_{st}\,z \qquad (3.10)$$

The point of action of F_c can be obtained in general (and not only for rectangular sections) by taking moments of all the elementary forces dF_c about, say, the neutral axis and equating this to $F_c\bar{x}$ (where $\bar{x}$ is the distance of the point of action from the neutral axis). That is,

$$F_c\bar{x} = {}_0\!\int^x dF_c \times y \qquad (3.11)$$

On combining with equation (3.6):

$$\bar{x} = {}_0\!\int^x (y/x)\,\sigma_c b y dy/(\tfrac{1}{2}\sigma_c bx) = \tfrac{2}{3}x \qquad (3.12)$$

Hence the lever arm is given by

$$z = \bar{x} + d - x = d - \tfrac{1}{3}x \qquad (3.13)$$

which, when introduced into equation (3.10), gives:

$$M_R = \tfrac{1}{2}\sigma_c bx(d - \tfrac{1}{3}x) = A_{st}\sigma_{st}(d - \tfrac{1}{3}x) \qquad (3.14)$$

It is often convenient to express the moment of resistance in the form $M_R = Kbd^2$, where K is a coefficient. For example, when the moment of resistance is based on concrete alone:

$$K = K_c = M_R/bd^2 = \tfrac{1}{2}\sigma_c bx(d - \tfrac{1}{3}x)/bd$$
$$= \tfrac{1}{2}\sigma_c na \qquad (3.15)$$

where a (= z/d) is the lever-arm factor. When based on steel:

$$K = K_{st} = M_R/bd^2 = \sigma_{st}A_{st}(d - \tfrac{1}{3}x)/bd^2$$
$$= \sigma_{st}\rho a \qquad (3.16)$$

Of course, for a balanced design $K_c = K_{st}$ and $\sigma_{st} = p_{st}$, while $\sigma_c = p_c$.

Example 3.1

A reinforced concrete beam (width = 280 mm) is required to carry a bending moment of 58.0 kN m.

(a) What depth of concrete is required?

(b) How many 25 mm diameter bars will be adequate for reinforcement?

Fig. 3.2 Cover to reinforcement in a beam

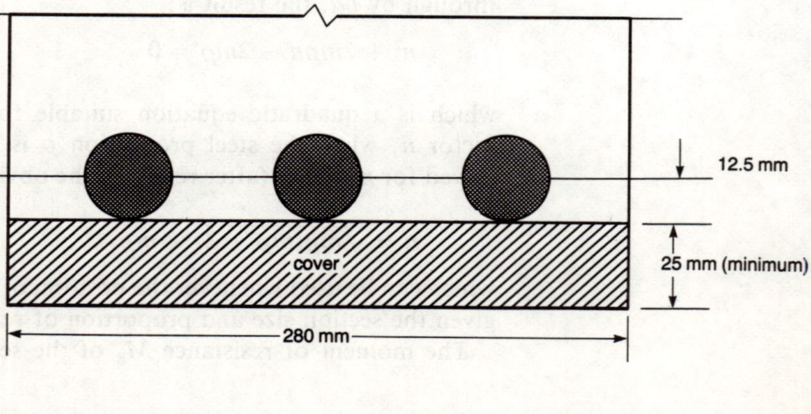

(mm)	1	2	3	4	5	6	7	8	9	10
6	28.3	56.5	84.8	113.1	141.4	169.6	197.9	226.2	254.5	282.7
8	50.3	100.5	150.8	201.1	251.3	301.6	351.9	402.1	452.4	502.7
10	78.5	157.1	235.6	314.2	392.7	471.2	549.8	628.3	706.9	785.4
12	113.1	226.2	339.3	452.4	565.5	678.6	791.7	904.8	1018	1131
16	201.1	402.1	603.2	804.2	1005	1206	1407	1608	1810	2011
20	314.2	628.3	942.5	1257	1571	1885	2199	2513	2827	3142
25	490.9	981.7	1473	1963	2454	2945	3436	3927	4418	4909
32	804.2	1608	2413	3217	4021	4825	5630	6434	7238	8042
40	1257	2513	3770	5026	6283	7540	8796	10 053	11 310	12 566

Table 3.1 Cross-sectional areas (mm^2) for a given number of bars of the same diameter

Assume the permissible stress in the concrete = 7 N/mm^2 and in the steel = 126 N/mm^2. Take the modular ratio m = 15.

Solution

Start by assuming a balanced section. From equation (3.4):

$$n = 1/(1 + p_{st}/mp_c) = 1/(1 + 126/15 \times 7) = 0.45$$

while, from equation (3.13):

$$a = z/d = 1 - n/3 = 0.85$$

Equation (3.15) gives:

$$K_c = \tfrac{1}{2} p_c na = \tfrac{1}{2} \times 7 \times (0.45) \times (0.85) = 1.34 \text{ N/mm}^2$$

By setting M_R = 58.0 kN m (at least) and as $K_c = M_R/bd^2$:

$$d = \sqrt{}(58 \times 10^6/1.34 \times 280) = 393 \text{ mm}$$

Clause 3.8(3) in NCP 1 stipulates that cover should be provided for a longitudinal reinforcing bar in a beam, and that it should not be less than 25 mm nor less than the diameter of the bar. The term cover used in the code implies the amount of concrete material actually covering the reinforcement (see exaggerated diagram Fig. 3.2).

As 25 mm diameter bars are to be used, the total 'embedment' (ie, nominal cover + radius of bar) cannot be less than 37.5 mm, say, 40 mm. Thus, the total depth required = 393 + 40 = 433 mm.

It is expedient to provide section sizes which are round figures (for obvious constructional convenience). The section size was therefore chosen as:

$$h = 440 \text{ mm}$$

The area of reinforcement required can be obtained from equation (3.14):

$$A_{st} = M_R/p_{st} \times z = 58 \times 10^6/126 \times 0.85\,(440-40) = 1354 \text{ mm}^2$$

Looking at the 25 mm diameter row in Table 3.1, which gives the ratio between area and number of bars, it can be seen that three 25 mm diameter bars (1473 mm^2) are required.

Example 3.2

A 225 mm × 450 mm rectangular beam is reinforced for tension with four 16 mm diameter bars. What are the maximum stresses in the concrete and the steel if the largest bending moment on the beam is 33.5 kN m? Take m = 15.

Solution Assuming a 40 mm embedment (or total cover):

$$d = 450 - 40 = 410\,\text{mm}$$

The area of reinforcement (from Table 3.1) is:

$$A_{st} = 804.2\ \text{mm}^2$$

The steel proportion:

$$\rho = 804.2/225 \times 410 = 8.7 \times 10^{-3}$$

and

$$m\rho = 15 \times 8.7 \times 10^{-3} = 0.13$$

Using equation (3.9) the neutral axis factor is:

$$n = \sqrt{0.13(0.13 + 2)} - 0.13 = 0.397$$

while the lever-arm factor, from equation (3.13), is given by:

$$a = 1 - n/3 = 0.868$$

Using equation (3.15):

$$K_c = \tfrac{1}{2}\,\sigma_c n a = M/bd^2$$
$$= 33.5 \times 10^6/225 \times 410^2$$
$$= 0.886\ \text{N/mm}^2$$

or

$$\sigma_c = 2 \times 0.886/0.397 \times 0.868$$
$$= \underline{5.1\ \text{N/mm}^2}$$

From equation (3.3):

$$\sigma_{st}/\sigma_c = m(1/n - 1)$$
$$= 15(1/0.397 - 1)$$
$$= 22.9$$
$$\therefore\ \sigma_{st} = 22.9 \times 5.1$$
$$= \underline{116.2\ \text{N/mm}^2}$$

Comments

The first example relates to the normal design of a section. Given the maximum applied moments, suitable section sizes and reinforcement are obtained starting from the premise of a balanced design. A large class of design problems falls into this category.

The second example involves the analysis of a section. Given the section size and the amount of reinforcement, the stresses developed under a specified maximum bending moment are computed. This second category of problem is usually met in practice when a designed structure (or one that is already built) is being checked for the stresses developed under ordinary working conditions. These stresses should not exceed the permissible stresses.

Both categories are important as far as the general principles of design are concerned.

3.1.2 Doubly-reinforced sections

When there is a restriction on the dimensions of a structural member, additional reinforcement may be introduced in the compression zone to make up for the reduction in the amount of concrete now available. Restrictions on the overall depth of a beam may arise out of a need for more head-room or for purely aesthetic reasons.

NCP 1 (clause 4.1.1.4) requires that, in rectangular beams, the amount of steel in compression should preferably not exceed 4 per cent of the total cross-sectional area but, if it does, 4 per cent should be allowed for in the calculation of the resistance moment of the beam.

A typical cross-section of a doubly-reinforced rectangular beam is shown in Fig. 3.3, along with the stress and strain curves across the section.

The area of concrete available to carry compression is $(bx - A_{sc})$. By proportion, the stress in the concrete at the location of the compression reinforcement is $(x - d')\sigma_c/x$. Therefore, the compression in the concrete is given by:

$$F_c = \tfrac{1}{2}\sigma_c bx - \sigma_c(x - d')A_{sc}/x \tag{3.17}$$

Furthermore, the strain at the location of the compression reinforcement is:

$$(x - d')\epsilon_c/x = (x - d')\sigma_c/E_c x$$

(as $\epsilon_c = \sigma_c/E_c$).

Therefore, the stress in the compression reinforcement is:

$$(x - d')\sigma_c E_{st}/E_c x = m\sigma_c(x - d')/x$$

Hence:

$$F_{sc} = m\sigma_c(x - d')A_{sc}/x \tag{3.18}$$

With $F_{st} = \sigma_{st} A_{st}$ the equilibrum equation $(F_{sc} + F_c = F_{st})$ becomes:

$$\tfrac{1}{2}\sigma_c bx - \sigma_c(x - d')A_{sc}/x + m\sigma_c(x - d')A_{sc}/x = \sigma_{st}A_{st} \tag{3.19}$$

or, by dividing through by σ_c and rearranging:

$$\tfrac{1}{2}bx + (m - 1)(x - d')A_{sc}/x = \sigma_{st}A_{st}/\sigma_c \tag{3.20}$$

Fig. 3.3 Doubly-reinforced rectangular section

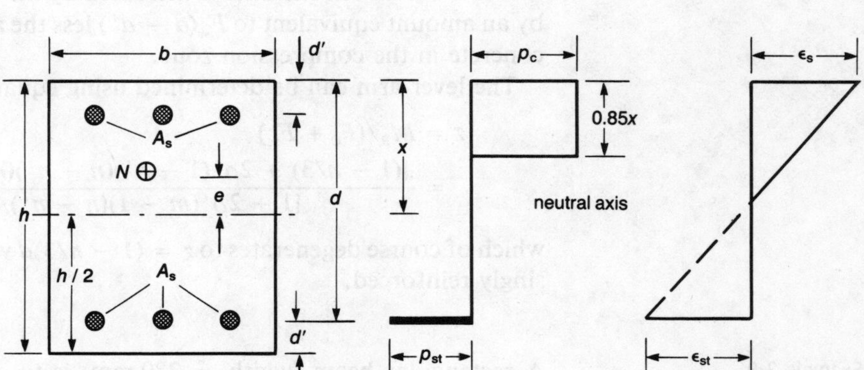

(a) Rectangular column section (b) Stress variation (c) Strain variation

From geometric consideration of the strain diagram (by comparing two similar triangles):

$$\sigma_{st}/\sigma_c = m(d - x)/x \tag{3.21}$$

which, when introduced into equation (3.20), gives:

$$\tfrac{1}{2}bx + (m - 1)(x - d')A_{sc}/x = m(d - x)A_{st}/x \tag{3.22}$$

By multiplying through by x/bd and setting $A_{st}/bd = \rho$, $A_{sc}/bd = \rho'$, $x/d = n$ and $d'/d = n'$, equation (3.22) becomes:

$$\tfrac{1}{2}n^2 + n\{(m - 1)\rho' + m\rho\} - \{(m - 1)n'\rho' + m\rho\} = 0 \tag{3.23}$$

which is a quadratic equation for the neutral axis factor, n. The positive root is the only admissible solution, giving:

$$n = -\{(m - 1)\rho' + m\rho\} + \sqrt{\{(m - 1)\rho' + m\rho\}^2 + 2\{(m - 1)n'\rho' + m\rho\}} \tag{3.24}$$

Equation (3.24) obviously degenerates to equation (3.9) for a singly-reinforced section when $\rho' = 0$.

The moment of resistance of the section based on the concrete is obtained by taking moments of all compressive forces about the location of the tension reinforcement. That is:

$$M_R = \tfrac{1}{2}\sigma_c bx(d - x/3) - A_{sc}(d - d')(x - d')\sigma_c/x + A_{sc}m\sigma_c(x - d')(d - d')/x \tag{3.25}$$

or:

$$K_c = M_R/bd^2 = \sigma_c\{\tfrac{1}{2}n(1 - n/3) + \rho'(1 - n')(m - 1)(n - n')/n \tag{3.26}$$

It is worth noting that the first term in equation (3.26), that is, $\tfrac{1}{2}\sigma_c n(1 - n/3)$, represents the moment-coefficient K_c for the singly-reinforced rectangular beam [compare equation (3.15), with $a = 1 - n/3$]. Hence the compression reinforcement has increased the factor K_c by an amount $\sigma_c\rho'(1 - n)(m - 1)(n - n')/n$. Thus a doubly-reinforced section may be viewed as consisting of an equivalent singly-reinforced section whose moment of resistance is increased by the compressive reinforcement, by an amount equivalent to $F_{sc}(d - d')$ less the amount due to the displaced concrete in the compression zone.

The lever-arm can be determined using equation (3.25) to give:

$$z = M_R/(F_c + F_{sc}) = \frac{\{(1 - n/3) + 2\rho'(1 - n')(n - n')(m - 1)/n^2\}d}{\{1 + 2\rho'(m - 1)(n - n')/n\}} \tag{3.27}$$

which of course degenerates to $z = (1 - n/3)d$ when $\rho' = 0$, ie, the beam is singly reinforced.

Example 3.3

A rectangular beam (width = 280 mm) is to carry a bending moment of 86.7 kN m. Given that the depth is to be limited to 460 mm, calculate the amount of reinforcement required. Take $p_c = 7\ \text{N/mm}^2$, $p_{st} = 126\ \text{N/mm}^2$ and $m = 15$.

Solution

For a typical design undertaken in the design office, the need for compression reinforcement may not be known beforehand.

Therefore, start by assuming the beam is to be reinforced for tension only. The effective depth $d = 460 - 40 = 420$ mm.

Assuming a balanced design:

$$n = 1/(1 + p_{st}/mp_c) = 0.45$$

and

$$a = 1 - n/3 = 0.85$$

$$x = nd = 0.45 \times 420 = 189 \text{ mm}$$

Also

$$K_c = \tfrac{1}{2} p_c na = 1.34 \text{ N/mm}^2$$

$$\therefore M_R = K_c bd^2 = 1.34 \times 280 \times 420^2$$

$$= 66\,185\,280 \text{ N mm}$$

But the applied moment = 86 700000 N mm, which is larger than the moment of resistance of the beam if it is singly-reinforced. Hence compression reinforcement is required. The extra moment to be carried by compression is:

$$86\,700\,000 - 66\,185\,280 = 20\,514\,720 \text{ N mm}$$

The effective stress on the compression reinforcement (see equation 3.20) is:

$$(m - 1)(x - d')p_c/x = 14 \times (189 - 40) \times 7/189$$

$$= 77.3 \text{ N/mm}^2$$

(Note that a total cover of 40 mm has been allowed for the compression reinforcement.)

The moment arm of the compression reinforcement (about the location of the tension reinforcement) is $420 - 40 = 380$ mm. Hence the moment to be carried by the compression reinforcement is given by:

$$(A_{sc} \times 77.3) \times 380 = 20\,514\,720$$

or

$$A_{sc} = 698 \text{ mm}^2$$

Equation (3.22) can now be used to compute A_{st}, that is:

$$A_{st} = \frac{\tfrac{1}{2}(280)(189^2) + (14)(189 - 40)(699)}{15(420 - 189)}$$

$$= 1864 \text{ mm}^2$$

Example 3.4

The trapezoidal section of a doubly-reinforced concrete beam is shown in Fig. 3.4. Given that A_{sc} = total area of compression reinforcement, A_{st} = total area of tension reinforcement and m = the modular ratio:

- derive a general expression for the depth of the neutral axis x as a function of b, α, d, d', h, A_{st}, A_{sc} and m.
- obtain an expression for the moment of resistance M_R of the section based on the concrete.

Fig. 3.4 Doubly-reinforced trapezoidal section

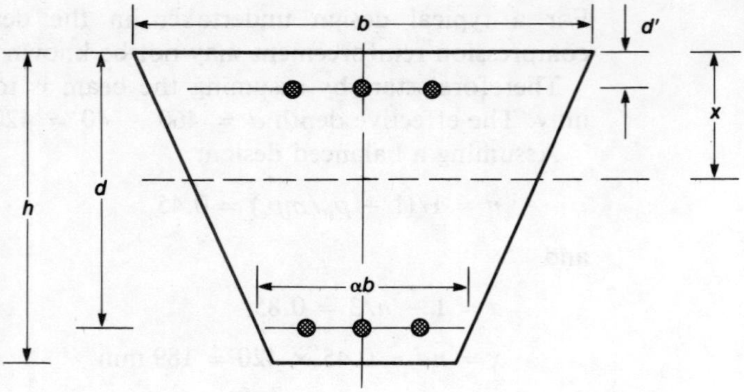

- from the results obtain the expression for the moment of resistance of a singly-reinforced trapezoidal section.

Solution

The stress variation across the section is drawn in perspective as a block (which it is) for clarity (Fig. 3.5). The volume of the block (in the compression zone) is essentially equal to the total compression developed in the concrete.

The stress in the concrete at the location of the compression reinforcement = $(x - d')\sigma_c/x$. By considering an element of the stress block (width = w, thickness = dy) at distance y from the neutral axis, the compression carried by the concrete alone can be determined as:

$$F_c = {}_0\int^x (w\mathrm{d}y)(y/x)\sigma_c - (x - d')\sigma_c A_{sc}/x \tag{3.28}$$

Fig. 3.5 Stress block for a trapezoidal section

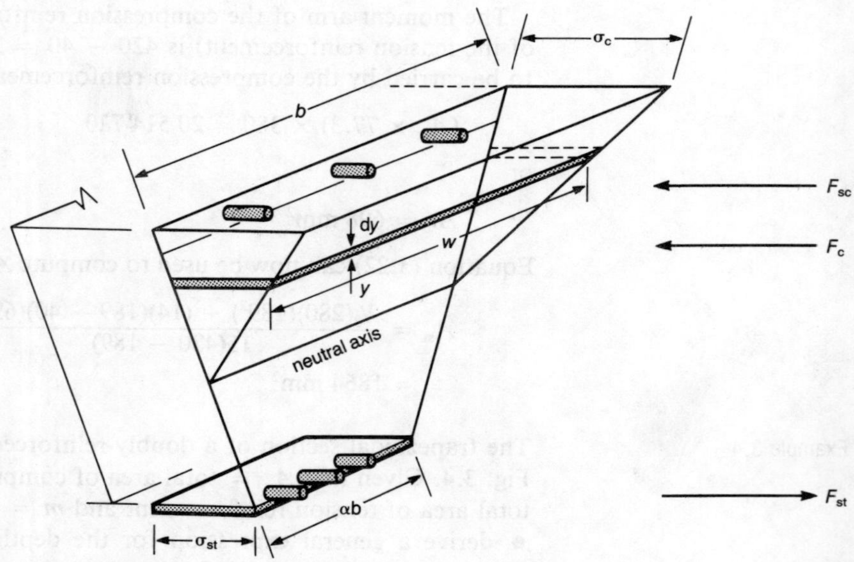

By proportion, $w = Ay + B$ where $A = b(1 - \alpha)/d$ and $B = b - Ax$. Thus:

$$F_c = (\sigma_c/x)\,{}_0\!\int^x (Ay + B)y\,dy - (x - d')\sigma_c A_{sc}/x$$
$$= \sigma_c x(2Ax + 3B)/6 - (x - d')\sigma_c A_{sc}/x \qquad (3.29)$$

The stress developed in the compression reinforcement $= m\sigma_c(x - d')/x$. Therefore, the compression carried by the compression reinforcement is:

$$F_{sc} = m\sigma_c(x - d')A_{sc}/x \qquad (3.30)$$

The force developed in the tension reinforcement is:

$$F_{st} = \sigma_{st}/A_{st} \qquad (3.31)$$

Using the equilibrium condition $F_{st} = F_c + F_{sc}$ it follows that:

$$(\sigma_{st}/\sigma_c)A_{st} = x(2Ax + 3B)/6 + (m - 1)(x - d')A_{sc}/x \qquad (3.32)$$

But from the strain distribution (equation 3.21) $\sigma_{st}/\sigma_c = m(d - x)/x$, which can be introduced into equation (3.32) to yield:

$$m(d - x)A_{st} = x^2\{(2Ax + 3B)/6 + (m - 1)(x - d')A_{sc}\} \qquad (3.33)$$

Putting $A = b(1 - \alpha)/d$, $B = b - Ax$ into equation (3.33) and re-arranging, the result is:

$$-\{(1 - \alpha)b/6d\}x^3 + (\tfrac{1}{2}b)x^2 + \{mA_{st} + (m - 1)A_{sc}\}x$$
$$- \{mdA_{st} + (m - 1)d'A_{sc}\} = 0 \qquad (3.34)$$

which is the required expression for the neutral axis depth x. A closed-form solution for x is difficult to obtain but if it is assumed that $(1 - \alpha)b/6d = \beta << 1$ (which will be true for most realistic trapezoidal sections), an approximate solution (using the method of small perturbations) is:

$$x = x_0 + \beta x_0{}^3/(\Omega_0 + bx_0) \qquad (3.35)$$

in which $x_0 = \{\sqrt{(\Omega_0{}^2 + \Omega_1)} - \Omega_0\}d$, $\Omega_0 = \{mA_{st} + (m - 1)A_{sc}\}/b$

and $\Omega_1 = 2\{mA_{st}\,d + (m - 1)d'A_{sc}\}/b$.

The moment of resistance can be obtained by taking moments about the location of the tension reinforcement, so that:

$$M_R = {}_0\!\int^x (w\,dy)(y/x)\sigma_c(y + d - x) + (x - d)\sigma_c A_{sc}(d - d')(m - 1)/x$$

With $w = Ay + B$ the above integration can be carried out (using some algebra) and the moment of resistance is then given by:

$$M_R = 1/12\,\sigma_c xb\,\{(1 - \alpha)(x - 2d)x/d + 6(d - x/3)\}$$
$$+ (m - 1)\,\sigma_c A_{sc}(x - d')(d - d')/x \qquad (3.36)$$

For a singly-reinforced trapezoidal beam $A_{sc} = 0$ and:

$$M_R = 1/12\,\sigma_c xb\,\{(1 - \alpha)(x - 2d)\,x/d + 6(d - x/3)\} \qquad (3.37)$$

which degenerates to $M_R = \tfrac{1}{2}\sigma_c bx(d - x/3)$ when $\alpha = 1$ or the beam is rectangular. Thus:

$$M_R\ \textbf{(rectangular)} - M_R\ \textbf{(trapezoidal)} =$$
$$1/12\,\sigma_c x^2 b(1 - \alpha)(2d - x)/d \qquad (3.38)$$

which is the reduction in resistance moment due to the 'bevelling' which takes place when a rectangular section becomes a trapezoidal one. Note that the result of equation (3.38) is independent of whether the beam is singly or doubly reinforced.

3.1.3 Flanged sections

Generally, slabs are constructed to be monolithic with beams. Therefore most beams should normally be treated as flanged members. Whenever a slab is monolithic with a beam there exists a region within the slab that aids the beam in carrying some of the compression. Usually a flanged beam is either L- or T-shaped (see Fig. 1.8) although a Z-shape may exist in certain structures with split-level floors.

The Nigerian Code of Practice NCP 1 (clause 4.1.1.5) requires that, in T-beams, the breadth of the flange assumed to be taking compression *should not* exceed whichever is the least of the following:
(a) one-third of the effective span of the T-beam;
(b) the distance between the centres of the ribs of the T-beams;
(c) the breadth of the rib plus twelve times the thickness of the slab.
The equivalent requirement for L-beams is contained in clause (4.1.1.6), which states that the flange assumed to be taking compression *should not* exceed whichever is the least of the following:
(a) one-sixth of the effective span of the L-beams;
(b) the breadth of the rib plus one-half of the clear distance between ribs;
(c) the breadth of the rib plus four times the thickness of the slab.
The first step in the design of a flanged beam is the location of the neutral axis. If the neutral axis falls below the soffit of the flange the designer has to decide whether to consider or ignore the compression in the part of the rib above the neutral axis. Conventionally it is thought that this compression can usually be neglected without appreciable effects on the overall results (see the example below and Exercise 3.12).

Fig. 3.6 Singly-reinforced T-section

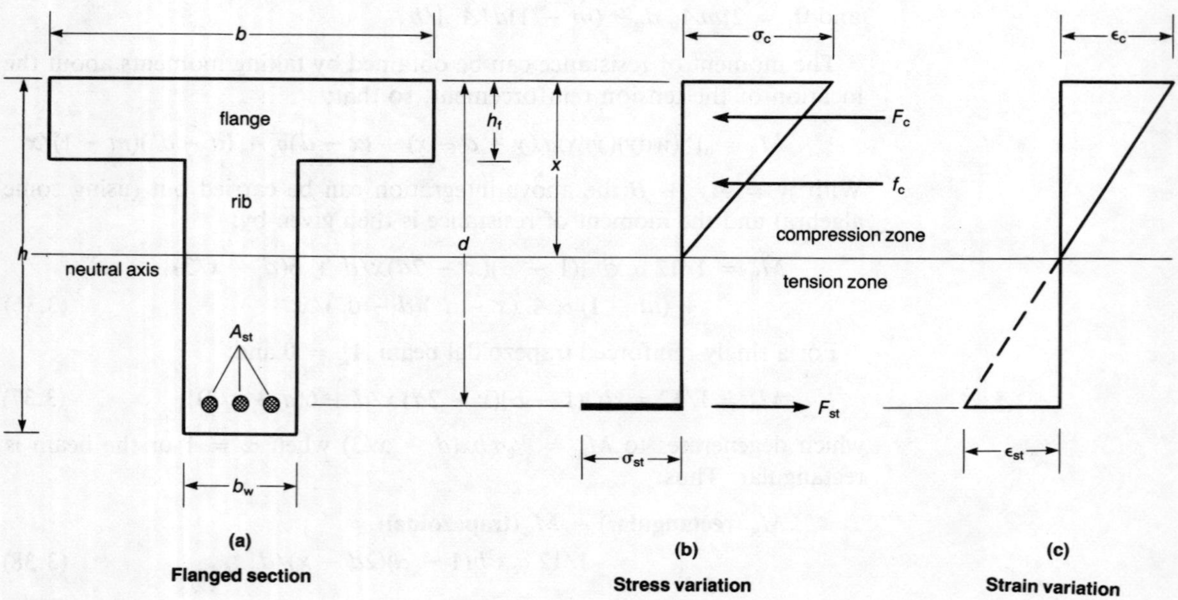

If the neutral axis falls within the flange, the beam behaves like a rectangular beam whose width is the least of (a), (b) or (c) above.

Consider the T-beam whose cross-section is shown in Fig. 3.6. The stress in the compression zone at a section distance y above the neutral axis is $(y/x)\sigma_c$. The compression on an element of thickness dy in the flange is $y\sigma_c b\mathrm{d}y/x$. As y varies from $x-h_f$ (at the soffit of the flange) to x (at the top of the flange), the total compression carried by the flange is given by

$$F_c = (b\sigma_c/x) \int_{x-h_f}^{x} y\mathrm{d}y$$

or

$$F_c = \tfrac{1}{2}h_f(2x - h_f)b\sigma_c/x \tag{3.39}$$

Similarly, the compression carried by the portion of the rib above the neutral axis is:

$$f_c = (b_w\sigma_c/x) \int_{0}^{x-h_f} y\mathrm{d}y$$

or

$$f_c = \tfrac{1}{2}b_w(x - h_f)^2\sigma_c/x \tag{3.40}$$

The tension in the reinforcement is $F_{st} = \sigma_{st}A_{st}$. For equilibrium $F_c + f_c = F_{st}$, or:

$$\tfrac{1}{2}h_f(2x - h_f)\, b\sigma_c/x + b_w(x - h_f)^2\sigma_c/x = \sigma_{st}A_{st} \tag{3.41}$$

From the strain diagram, $\sigma_{st}/\sigma_c = m(d - x)/x$ which, when introduced in equation (3.41) and after rearranging, yields:

$$(\tfrac{1}{2}b_w)x^2 + \{(b - b_w)h_f + mA_{st}\}x - \tfrac{1}{2}\{(b - b_w)h_f^2 + 2mdA_{st}\} = 0 \tag{3.42}$$

which is a quadratic equation that can be used to determine the neutral axis depth x. After rejecting the unrealistic negative root, the solution for x is in the form:

$$x = -\Omega_0 + \sqrt{(\Omega_1 + \Omega_0^2)} \tag{3.43}$$

in which

$$\Omega_0 = \{(b - b_w)h_f + mA_{st}\}/b_w$$

and

$$\Omega_1 = \{(b - b_w)h_f^2 + 2mdA_{st}\}/b_w$$

If the rib forces are neglected (that is, $f_c << F_c$) and with $\sigma_{sc}/\sigma_c = m(d-x)/x$ equation (3.41) reduces to:

$$\tfrac{1}{2}h_f(2x - h_f)b\sigma_c/x = m(d - x)A_{st}/x$$

or

$$x = \frac{\tfrac{1}{2}bh_f^2 + mdA_{st}}{bh_f + mA_{st}} \tag{3.44}$$

which is a simple expression commonly used in the design office to obtain the neutral axis depth.

Taking moments about the location of the tension reinforcement, the moment of resistance (based on the concrete) is given by:

$$M_R = (b\sigma_c/x)\int_{x-h_f}^{x} y(y + d - x)\mathrm{d}y + (b_w\sigma_c/x)\int_{0}^{x-h_f} y(y + d - x)\mathrm{d}y \quad (3.45)$$

or

$$M_R = \tfrac{1}{2}bx\sigma_c(d - x/3) - (b - b_w)(x - h_f)^2(d - \tfrac{2}{3}x - h_f)\sigma_c/x \quad (3.46)$$

The lever-arm is:

$$z = M_R/(F_c + f_c)$$

$$z = \frac{bx^2(d - x/3) - 2(b - b_w)(x - h_f)^2(d - \tfrac{2}{3}x - h_f)}{bh_f(2x - h_f) + b_w(x - h_f)^2}$$

which reduces to $z = d - x/3$ if the neutral axis is chosen to coincide with the soffit of the flange (ie, $x = h_f$). The moment of resistance based on the reinforcement is simply $M_R = \sigma_{st}A_{st}\,z$.

Example 3.5

A T-beam has a rib width of 230mm, a flange thickness of 150mm and an overall depth of 460mm. If the flange width is 920mm and the area of reinforcement is 3870mm² determine:

(a) the neutral axis depth by taking the rib forces into consideration;
(b) the neutral axis depth by ignoring the rib forces; and
(c) from the results of (a) above find the maximum stress in the concrete, given that the stress in the steel is 140N/mm². Take $m = 15$.

Solution

(See also Fig. 3.7)

(a) To locate the position of the neutral axis use equation (3.43) in which:

$$\Omega_0 = \{(920 - 230)150 + 15(3870)\}/230 = 702.39\,\text{mm}$$

$$\Omega_1 = \{(920 - 230)150^2 + 2(15)(420)(3870)\}/230 = 2.80 \times 10^5\,\text{mm}^2$$

$$\therefore x = -\Omega_0 + \sqrt{(\Omega_1 + \Omega_0^2)} = -702.39 + \sqrt{(2.80 \times 10^5 + 702.39^2)} = 176.7\,\text{mm}$$

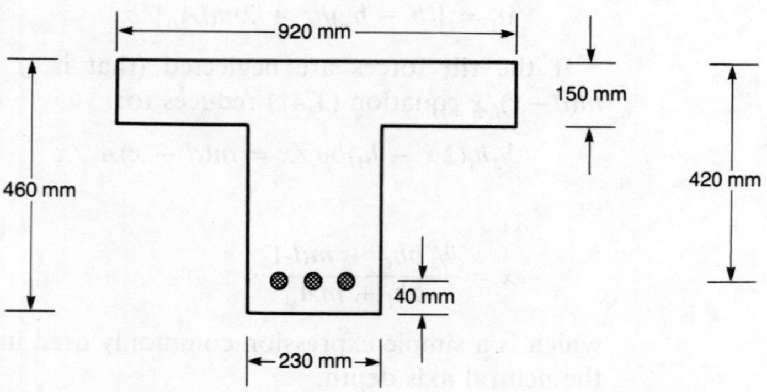

Fig. 3.7 Solution to Example 3.5(a)

(b) If the rib forces are neglected equation (3.44) can be used to give:

$$x = \frac{\frac{1}{2}(920)150^2 + 15(420)(3870)}{920(150) + 15(3870)}$$

$$= 177.2\,\text{mm}$$

which shows that ignoring the rib forces has produced an error (in the neutral-axis depth x) equal to $(176.7 - 177.2)/176) \times 100 = -0.3$ per cent. (See Exercise 3.12 for the corresponding error in the estimation of the resistance moment if the rib forces are neglected.)

(c) With $x = 176.7$ mm the use of the strain diagram results in:

$$\sigma_{st}/\sigma_c = m(d - x)/x = 15(420 - 176.7)/176.7$$

$$= 20.7$$

or

$$\sigma_c = 140/20.6 = 6.8\ \text{N/mm}^2$$

3.2 Deflection

As stated earlier, with the elastic method serviceability requirements are met indirectly via the satisfaction of strength requirements. However, in order to avoid excessive deflections NCP 1 (clause 3.10.2) has placed limits on the span/depth ratios for various support conditions. The ratios are reproduced in Table 3.2.

Table 3.2 Maximum span/depth ratios for beams

Beam types	Support conditions			
	simply supported	one end continuous	both ends continuous	cantilever
	20	23	26	10

Generally, in the design of beams the depth is first chosen to satisfy the bending requirements. The span/depth ratio is then computed and checked against the limits stipulated in Table 3.2. If the span/depth ratio exceeds the maximum allowed, deflection is used as the design criterion and the minimum depth required will be based on the ratios in the above table.

3.3 Shear

The development of shear stresses in a reinforced concrete member is a complicated process which has not yet been fully understood. Simple analytical expressions can be obtained for the stress distribution in structures whose materials are homogeneous and elastic. However, these expressions are not applicable to reinforced concrete, mainly because it is not homogeneous. Therefore the shear design equations currently in use are generally based on statistical analysis of experimental data.

For beams it is now known that certain 'diagonal tension stresses', developed at the instance of shear stresses, are invariably responsible for

Fig. 3.8 Shear reinforcement in beams

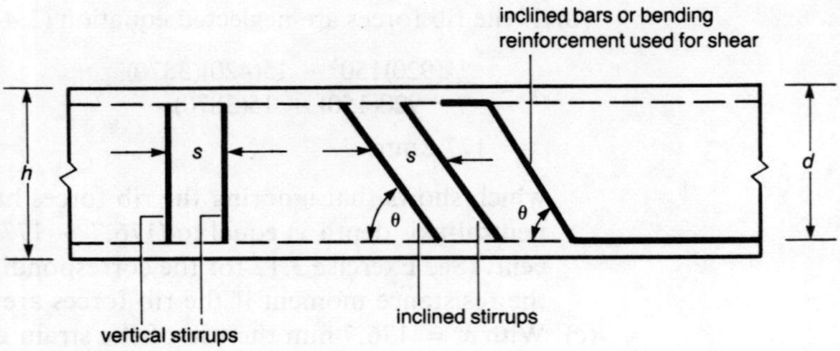

failure in the concrete (rather than the shear stresses). In any case the nominal shear stress is used as a measure of the diagonal tension.

NCP 1 (clause 3.12.1) requires that the shear stress v at any cross-section in a reinforced concrete beam (or slab) be calculated from the expression:

$$v = V/bz \tag{3.48}$$

in which V is the total shearing force across the section, b is the breadth of a rectangular beam (which for a T-beam or an L-beam should be replaced by the breadth of the rib b_w) and z is, of course, the lever arm.

If the shear stress calculated from equation (3.48) exceeds the permissible shear stress in the concrete, provision must be made to aid the concrete in carrying the shear forces (actually, the diagonal tension stresses).

The permissible shear stresses for nominal concrete mixes are presented in Table 2.1. The permissible shear stress is about one-tenth of the permissible compressive stress in bending. In no case should the shear stress, as calculated from equation (3.48), exceed four times the permissible shear stress for the concrete alone.

Shear reinforcement in beams consists basically of two types (Fig. 3.8):

- 'stirrups' or 'links' which may be placed perpendicularly to the longitudinal reinforcement or at some angle to the horizontal; or
- reinforcement used to resist bending moments bent diagonally upward and making an angle of at least 30° with the axis of the longitudinal portion of the bar.

The spacing of shear stirrups should not exceed the lever arm z of the beam. The spacing s required for a given shear force can be computed from the expression:

$$s = p_{st}A_w z/V \tag{3.49}$$

in which p_{st} is the permissible tensile stress in the shear reinforcement and A_w is the cross-sectional area of the stirrups.

When inclined bars are used, the resistance to shear at any section may be calculated on the assumption that the inclined bars form the tension members of one or more single systems of lattice girders, in which the concrete forms the compression members. The shear resistance of any vertical section is taken as the sum of the vertical components of the tension and compression forces cut by the section. It is important to ensure that these assumptions do not lead to stresses in the horizontal bars that would be greater than the permissible stresses.

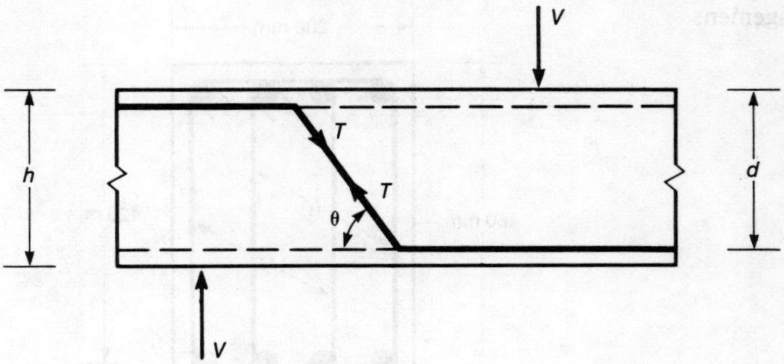

Fig. 3.9 Inclined bars used for shear reinforcement

In Fig. 3.9, if V = the shear force at a given section then the tension T in the inclined bars is given by balancing vertical forces. That is:

$$T \sin \theta = V$$

or

$$T = V/\sin \theta \tag{3.50}$$

It is necessary to ensure that:

$$\sigma_{st} = T/A_w \not> p_{st} \tag{3.51}$$

Example 3.6

A 280 mm × 460 mm rectangular beam is required to carry a maximum shear force of 99.4 kN. Determine the shear stress and (if shear reinforcement is required) the spacing of 10 mm diameter stirrups if:

(a) the stirrups are to be arranged perpendicularly to the longitudinal axis of the beam; or

(b) the stirrups are inclined at an angle of 45°. Assume p_{st} = 126 N/mm², p_c = 7 N/mm² and m = 15.

Solution

From equation (3.48) shear stress $v = V/bz$. Assuming a balanced section $n = 0.45$, $a = 0.85$ and $z = 0.85(460 - 40) = 357$ mm, then:

$$v = 99.4 \times 10^3/280 \times 357 = 0.99 \text{ N/mm}^2$$

But the permissible shear stress = 7/10 = 0.7 N/mm (see also Table 2.1). Therefore shear reinforcement is required.

(a) If four legs of 10 mm stirrups are chosen the spacing, using equation (3.49), is:

$$s = \frac{126 \times (4 \times \pi \times 10^2/4) \times 357}{99.4 \times 1000} = 142 \text{ mm}$$

Therefore, use four legs 10 mm diameter stirrups at 140 mm spacing.

Comments

The choice of the number of legs is a prerogative of the designer and may have to be reviewed if the resulting spacing is so small that it is impractical or so large that it exceeds the lever arm z. The arrangement of the four legs can be achieved as shown in Fig. 3.10.

Fig. 3.10 Arrangement of shear stirrups

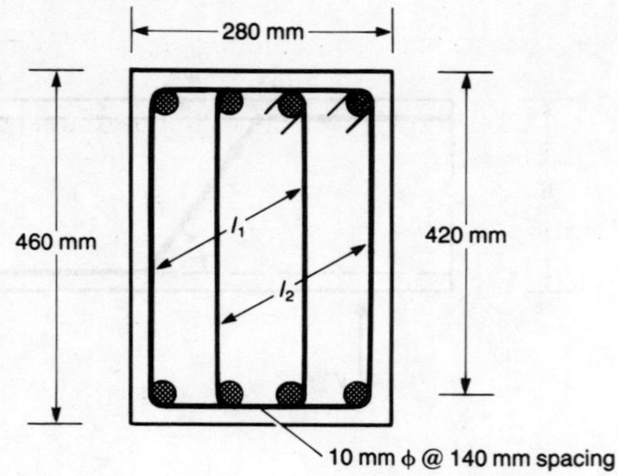

(b) If the stirrups are inclined at an angle of 45° then the total force to be carried by the stirrups is [see equation (3.50)]:

$T = 99.4/\sin 45° = 140.6$ kN

Choosing six legs (or three pairs of double links), the spacing is:

$$s = \frac{126 \times (6 \times \pi \times 10^2/4 \times 357)}{140.6 \times 1000} = 151 \text{ mm}$$

Therefore, use six legs 10 mm diameter stirrups at 150 mm spacing.

3.4 Torsion

The Nigerian Code of Practice NCP 1, and also the parent code CP 114, does not contain specific guidelines for the design for torsion. The major problem is that the complete behaviour of reinforced concrete members under torsion (especially with regard to the interaction with bending moments and shear) is not fully understood.

What is known is that failure of a concrete beam due to torsional moment results in diagonal cracks that run in an almost helical manner around the beam, along the longitudinal axis. The British Code of Practice CP 110, taking advantage of research efforts predating 1972, does have specific recommendations for torsion (see section 9.5).

The reinforcement for torsion normally consists of stirrups arranged perpendicularly to the longitudinal axis of the beam. These links, together with the longitudinal reinforcement, can be arranged to provide the required torsional moment of resistance.

One formal approach to the calculation is to use the space truss analogy due to Lampert and Collins (1972). If the beam is torsionally under-reinforced the links and longitudinal reinforcement will attain their respective permissible stresses p_t, p_{st} and the torsional moment of resistance T for a rectangular beam will be given bv

$$T = x_1 y_1 \sqrt{\left\{\frac{(A_t p_t)(A_{st} p_{st})}{(x_1 + y_1)s}\right\}} \qquad (3.52)$$

Where A_{st} is the area of the longitudinal reinforcement; A_t is the area of the two legs of each link; s is the longitudinal spacing of the links; and x_1 and y_1 are, respectively, the smaller and larger dimensions of the links.

3.5 Exercises

(Unless otherwise stated take $p_c = 7\ \text{N/mm}^2$, $p_{st} = 126\ \text{N/mm}^2$ and $m = 15$)

3.1 A rectangular beam is subjected to a maximum bending moment of 75.5 kN m. How many 20 mm diameter bars are required, given that the overall depth is twice the width? Determine a suitable section size.

3.2 A square beam is to carry a bending moment of 90.0 kN m. Find a suitable section size and the appropriate area of mild steel reinforcement.

3.3 A rectangular beam (6 m long and 225 mm wide) rests on block walls at each end. The beam carries a uniformly-distributed load of 10.5 kN/m and a point load of 52.0 kN acts through the centre.

(a) What is the minimum depth required from bending considerations?

(b) Will the computed depth satisfy the requirements of deflection?

(c) How many 25 mm diameter bars are needed for flexural requirements?

Fig. Q3.4

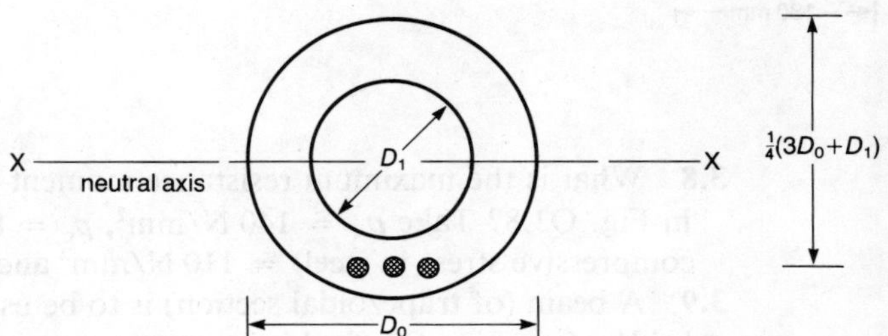

3.4 A singly-reinforced hollow circular beam (Fig. Q3.4) is to be designed such that the neutral axis coincides with the horizontal plane *x-x* passing through the centre of the beam.

(a) Obtain an expression for the area of reinforcement A_{st} required as a function of D_1 and D_0.

(b) By working from first principles determine the moment of resistance of the section based on the concrete alone.

(c) From the results of (b) determine the lever arm of a thin-walled, singly-reinforced circular concrete beam whose neutral axis coincides with the diameter. [**Hint**: $(D_0 - D_1)/D = \epsilon << 1$]

3.5 How many 25 mm diameter steel bars are needed to reinforce a hollow circular beam that is to carry a bending moment of 203.3 kN m, given that:

- the exterior diameter = 1000 mm;
- the interior diameter = 600 mm;
- $p_{st} = 140\ \text{N/mm}^2$, $p_c = 7\ \text{N/mm}^2$ and $m = 15$.

(**Hint**: The neutral axis does not necessarily coincide with the diameter).

3.6 A rectangular beam is to carry a uniformly-distributed load of 18.0 kN/m. The beam is fixed to a column at one end and is left free at the other. If the beam is 250 mm wide and the effective span is 2.0 m:

(a) obtain a suitable depth;

(b) calculate the area of reinforcement required;

(c) sketch the designed section and indicate the position of the reinforcing bars.

3.7 Repeat Question 3.6(b), assuming that the free end now rests on a block wall.

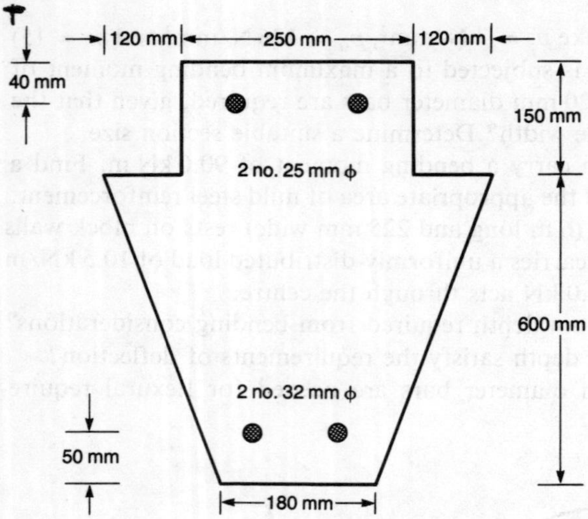

Fig. Q3.8

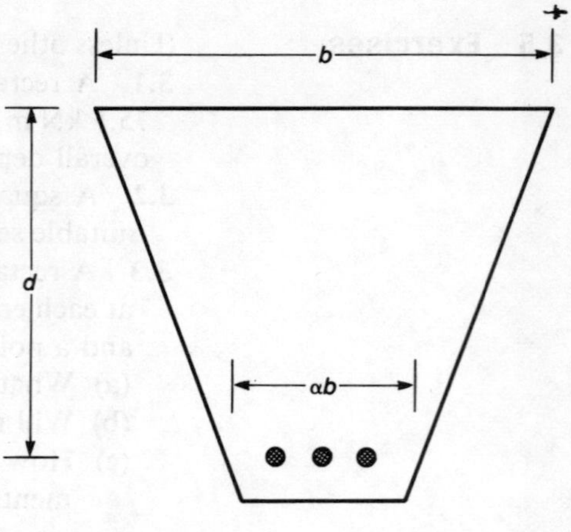

Fig. Q3.9

3.8 What is the maximum resistance moment of the beam section shown in Fig. Q3.8? Take p_{st} = 140 N/mm^2, p_c = 8.5 N/mm^2, p_{sc} (permissible compressive stress in steel) = 110 N/mm^2 and m = 15.

3.9 A beam (of trapezoidal section) is to be used to support the deck of a bridge (see Fig. Q3.9). If the maximum applied moment is M units, obtain:

(a) an expression for the area A_{st} of steel required for reinforcement;

(b) the ratio $\beta = A_{st}/A'_{st}$, where A'_{st} is the area of reinforcement required for an equivalent rectangular section (width b, effective depth d) carrying an applied moment of M units.

(c) the number of 32 mm diameter bars required if M = 500 kN m, b = 400 mm, d = 700 mm and α = 0.8. Take p_{st} = 140 N/mm^2, p_c = 7 N/mm^2 and m = 15.

3.10 The moment of resistance of a given rectangular beam (breadth b, effective depth d) is given by the expression $M_R = K_c bd^2$, where $K_c = \frac{1}{2} p_c n a$, $a = 1 - n/3$, and n is the neutral axis factor. If a moment M is applied on the beam section, the area of reinforcement A_{st} required can be obtained by setting $M = A_{st}\, p_{st}\, ad$. Using these expressions:

(a) obtain a relationship between M/M_R and A_{st}/bd;

(b) plot a graph of M/M_R versus A_{st}/bd for varying values of p_{st}/p_c between 10 and 30 (assuming m = 15);

(c) from the graph obtain the area of reinforcement required for a reinforced concrete beam, for which b = 230 mm, d = 410 mm, M = 70.0 kN m, p_{st} = 126 N/mm^2, p_c = 7 N/mm^2, and m = 15.

3.11 A rectangular beam of width 690 mm is needed for the strip foundation of a residential building. The beam is subjected to a maximum bending moment of 303.7 kN m. Given that the depth of the beam is not to exceed 510 mm, determine the amount of reinforcement required.

3.12 With reference to the T-beam section shown in Fig. Q3.12 determine the percentage error in the computation of the moment of resistance when the rib forces are neglected. What are the conditions under which this error will be less than 1 per cent?

Fig. Q3.12

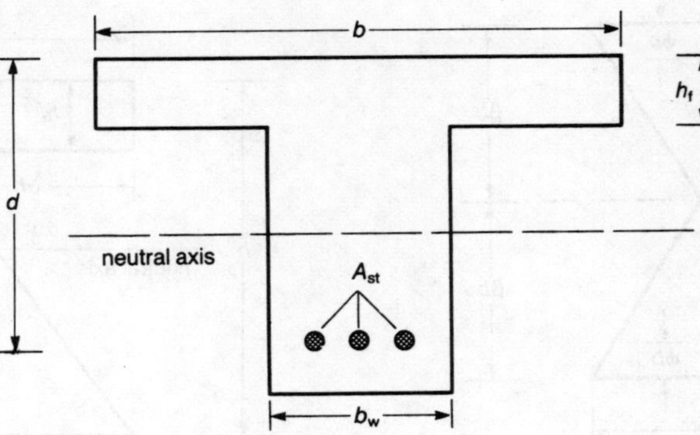

3.13 Obtain expressions for the neutral axis depths (where necessary) and the resistance moments of the various sections given in Figs. Q3.13(a) to (h). By taking appropriate limiting conditions, obtain the pertinent expressions for the resistance moments of some other special section shapes. From the results endeavour to plot dimensionless graphs that can readily be used in the design of these sections.

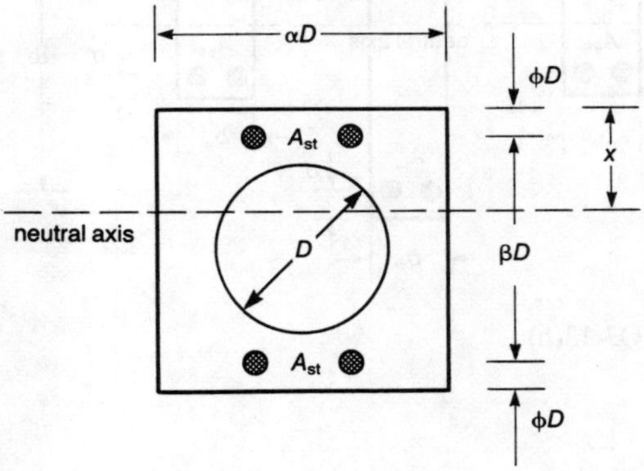

Fig. Q3.13(a)

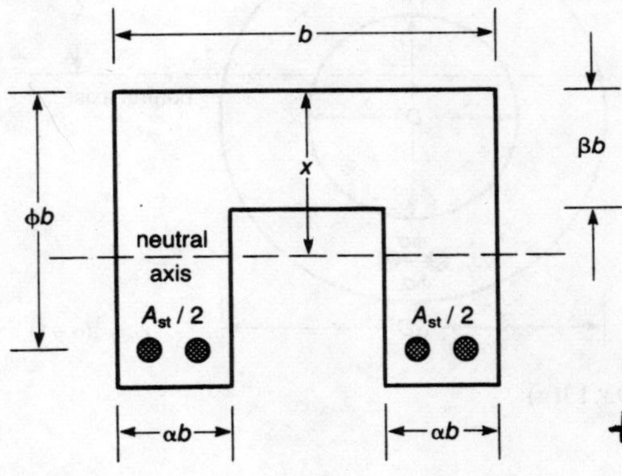

Fig. Q3.13(b)

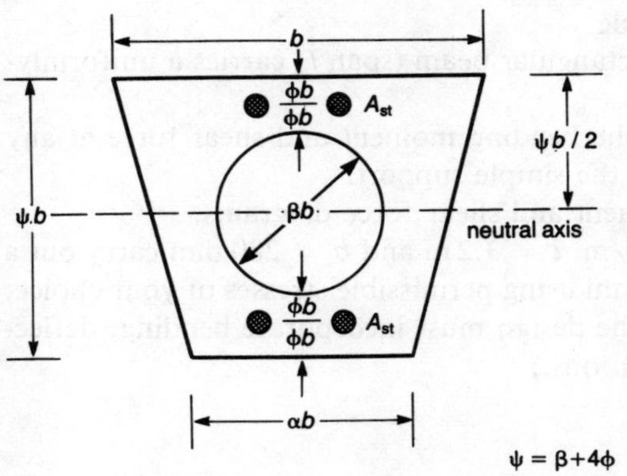

Fig. Q3.13(c)

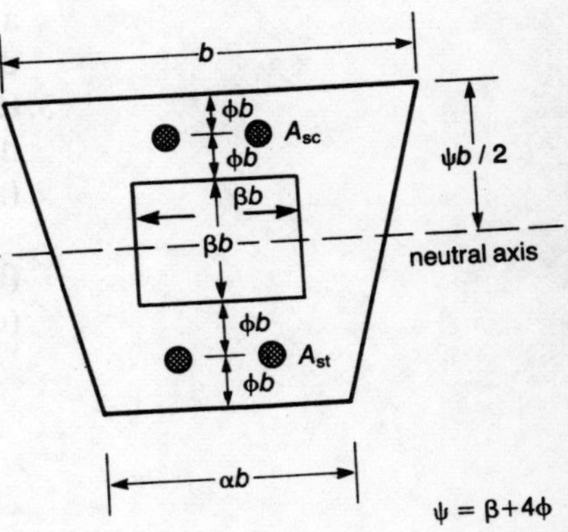

Fig. Q3.13(d)

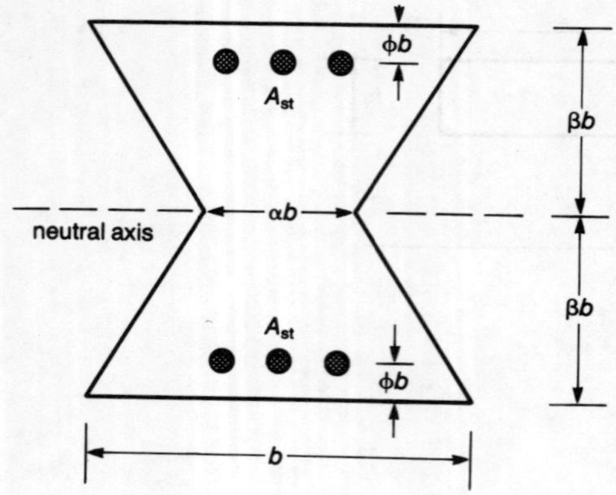

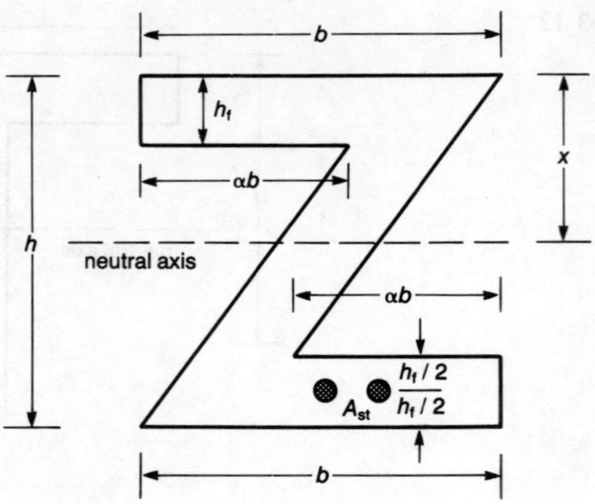

Fig. Q3.13(f)

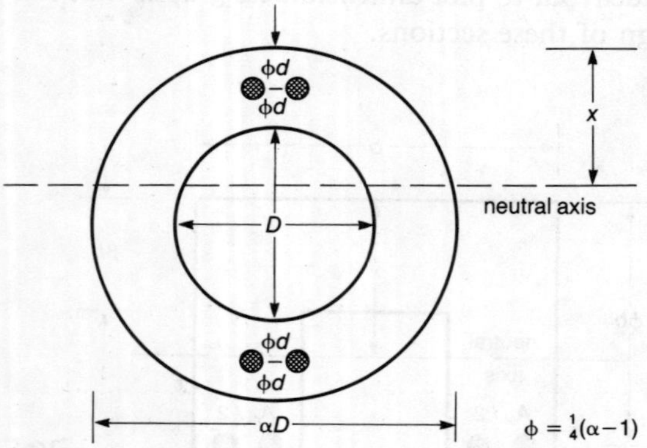

Fig. Q3.13(g)

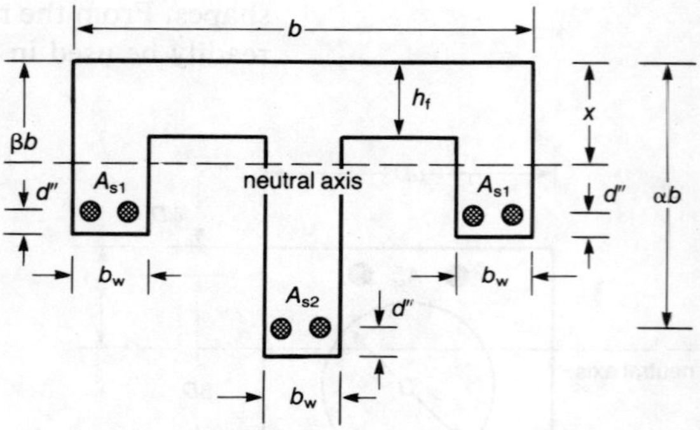

Fig. Q3.13(h)

3.14 Draw the diagram of shear forces for the beam given in Exercise 3.3 and then perform a detailed design for shear, assuming that only 8 mm diameter stirrups are available.

3.15 A propped cantilever rectangular beam (span l) carries a uniformly-distributed load of w kN/m.

(a) Obtain expressions for the bending moment and shear force at any section, distance y from the simple support.

(b) Sketch the bending moment and shear force diagrams.

(c) Given that $w = 24.4$ kN/m, $l = 3.2$ m and $b = 280$ mm carry out a detailed design of the beam using permissible stresses of your choice. Take $m = 15$. (**Hint**: The design must incorporate bending, deflection and shear considerations.)

4 Slabs

4.1 Assumptions

In general, the design of a reinforced concrete slab follows the same pattern established in the preceeding chapter for the reinforced concrete beam. Slabs are generally used for floors and roofs. Most slabs (solid slabs) are of uniform thickness but there are some special types (eg, ribbed, voided and hollow-block slabs) which consist of a series of T-beams and/or voids (see section 1.2.1 for details).

Slabs that are supported along two opposite edges are said to span in one direction. When a slab runs over two or more supports it is referred to as being continuous over supports. A slab is said to span in two directions when it is supported on all four of its edges.

The implicit assumption in the foregoing is that a given slab is rectangular in plan, which is indeed true for most slabs. However, other special shapes occurring in practice include the oval, the circular and the trapezoidal and these are often used for the floors of executive offices, luxury hotels and similar prestige buildings.

4.2 Loading

In the design of slabs the following loading must be accounted for:

- dead load (which depends on the slab thickness);
- load due to floor finishes (eg, tiles, carpets, rugs, etc);
- load due to partitions;
- imposed or live load.

For the purpose of computing the dead load the density of plain concrete can be taken to be 2300 kg/m^3, while that of reinforced concrete can be taken to be 2400 kg/m^3 (unless better estimates can be obtained).

A brief summary of the recommended minimum uniformly-distributed imposed load on slabs (and roofs) is presented in Table 4.1 overleaf.

4.3 Slabs spanning in one direction

These are, perhaps, the easiest slabs to design although in practice they may not be the most common. When the slabs are supported by brick walls or are non-monolithic with supporting reinforced concrete beams, the bending moments are obtained in the same way as for simply-supported members.

In such cases, for uniformly-distributed loading, the maximum bending moment is $wl^2/8$, where w is the uniformly-distributed load and l is the span along the main direction (ie, the direction along which the slab is spanned).

If the slab is constructed monolithically with the supporting beam or if the ends are built into walls, allowance must be made for the hogging

Table 4.1 Minimum imposed load on slabs (and roofs)

Type of structure	Floor use/Comments	Uniformly-distributed load (kN/m²)
Assembly buildings	With fixed seating	4.0
	Without fixed seating	5.0
Bedrooms	Domestic buildings	1.5
	Hotels	2.0
	Institutions	1.5
Libraries	Without book storage	2.5 (for each metre of
	With book storage	4.0 storage height)
Colleges	Classrooms	3.0
	Dining rooms	2.0
	Dormitories	1.5
	Gymnasia	5.0
Stationery stores		4.0 (for each metre of storage height)
Factories		7.5
Garages	For vehicles < 250 kg	2.5
	Repair workshops and parking areas for vehicles > 250 kg	5.0
Hospitals	Wards, utility rooms operating theatres	2.0
Offices	Computer rooms, etc.	3.5
	Filing and storage	5.0
	General offices	2.5
Hotels	Bars, etc.	5.0
	Kitchens	3.0
Stairs	Houses < three storeys	1.5
	Others: loading same as the floors to which they give access	3.0–5.0
Corridors and foot bridges between buildings	Subject to crowd loading	4.0
	Subject to loads greater than the crowds	5.0
Roofs	Slopes < 10° and if accessible other than for clearing and repairs	1.5
	Slopes < 10° but not accessible except for clearing and repairs	0.75
	Slopes > 10° but < 30°	0.75
	Slopes > 30° but < 75°	$(75-\theta)/60$ (θ = roof slope)
	Slopes > 75°	0.0

moments developed at the edges, although the maximum sagging moments can still be calculated on the same basis as for the freely-supported type. For the case of uniformly-distributed loading the allowance for hogging moments is $wl^2/24$.

To design a slab a representative strip of width 1000 mm along the main direction is chosen. The design procedure for this rectangular strip is similar to that for rectangular beams given in chapter 3. The reinforcement obtained for the strip represents the main steel (or reinforcement) required along the main direction. A proportion of the main reinforcement (25 to 50 per cent) must be provided as reinforcement along the minor direction (ie, the direction perpendicular to the main direction). These minor reinforcement bars are commonly referred to as distribution steel. (See section 13.4 for details of how the main and distribution bars are arranged in slabs.)

When a rectangular slab that is supported on all of its four edges is being spanned in one direction (a prerogative of the designer) the shorter side constitutes the main direction.

Example 4.1

A reinforced concrete slab is supported on two simple supports 3 m apart. If the slab carries a uniformly-distributed load of 10 kN/m², determine the suitable slab thickness and reinforcement. Assume $p_c = 7\,\text{N/mm}^2$, $p_{st} = 126\,\text{N/mm}^2$ and $m = 15$.

Solution

Maximum bending moment $= 10 \times 10^3\,(3)^2/8 = 11\,250$ N m. Assuming a balanced section, $n = 0.45$, $a = 0.85$, and $K = 1.34\,\text{N/mm}^2$.

$$\therefore M_R = 1.34bd^2$$

If $M_R = 11.25 \times 10^6$ N mm and $b = 1000$ mm

$$d = \sqrt{\frac{11.25 \times 10^6}{1.34 \times 10^3}} = 91.6\text{ mm}$$

Clause 3.8(5) of the Nigerian Code of Practice NCP 1 requires that cover for reinforcement in a slab should not be less than 15 mm nor less than the diameter of such reinforcement.

Adopting a nominal cover of 15 mm and assuming 12 mm diameter bars are to be used, the total cover $= 15 + 6 = 21$ mm. Thus:

$$91.6 + 21 = 111.6\text{ mm}$$

Therefore a suitable thickness for the slab would be 120 mm.

The area of reinforcement required is given by:

$$A_{st} = \frac{M}{p_{st}z} = \frac{11.5 \times 10^6}{126 \times 0.85 \times (120 - 21)}$$
$$= 1061\text{ mm}^2$$

Table 4.2 can be used to select the appropriate spacing of 12 mm diameter bars, giving 12 mm diameter bars at 100 mm centre-to-centre which is equivalent to an area of 1131 mm² for a strip 1000 mm wide. This recommendation is customarily presented in the short form:

Use 12 mm ϕ bars @ 100 mm c/c (1131 mm²).

Area of distribution bars $= \frac{1}{4}(1061) = 265$ mm² which, from Table 4.2, requires 6 mm ϕ bars @ 100 mm c/c (283 mm²) for distribution.

Table 4.2 Bar areas for various bar spacings (mm)

Spacing (mm)										
Bar diameter (mm)	75	100	125	150	175	200	225	250	275	300
6	377	283	226	188	162	141	126	113	103	94
8	670	503	402	335	287	251	223	201	183	168
10	1047	785	628	524	449	393	349	314	286	262
12	1508	1131	905	754	646	565	503	452	411	377
16	2681	2011	1608	1340	1149	1005	894	804	731	670
20	4189	3142	2513	2094	1796	1571	1396	1257	1142	1047
25	6545	4909	3927	3272	2805	2454	2182	1963	1785	1636
32	—	8042	6434	5362	4596	4021	3574	3217	2925	2681
40	—	—	10 053	8378	7181	6283	5585	5027	4570	4189

*Alternatively the spacing can be obtained using the formula $s = 250 \times \pi \times \phi^2 / A_{st}$, in which ϕ is bar diameter in mm, A_{st} is the total area of reinforcement in mm^2 and s is the spacing in mm.

4.4 Slabs spanning in two directions

If the dimensions of a slab are such that the ratio of the longer side (l_y) to the shorter side (l_x) does not exceed 3, the slab may be spanned economically in two directions (provided, of course, it is actually supported on all four sides). NCP 1 (clause 4.1.5) recommends any one of three possible methods for designing slabs spanning in two directions. These are described below.

Method 1: This is a purely theoretical analysis based on the elastic theory. The bending moments in the slabs and beams may be calculated on the assumption that the slabs act as perfectly elastic thin plates. The moments of resistance of the slab sections should be calculated by the commonly-used elastic theory with $m = 15$.

The use of this method invariably involves the need to solve some fourth-order (bi-harmonic) partial differential equations whose unknowns are the displacements and certain artificial stress functions. The interesting feature of the stress functions is that when differentiated they yield the actual stresses acting on the plate (or slab).

For simple boundary geometries (eg, rectangular boundaries) with ordinary boundary conditions, closed-form solutions are usually available. For others, any one of the existing numerical techniques (eg, the finite difference, the finite element or the boundary element methods) for solving partial differential equations can be used.

Method 2: In this method the assessment of bending moments in the slabs is based on theoretical analysis, amplified and adjusted after taking into account experimental data. The moment of resistance can be based on the elastic theory with $m = 15$. Three recommendations are high-lighted under this method and these are outlined below:

Table 4.3 Bending moment coefficients for slabs spanning in two directions at right angles and simply supported on four sides

l_x / l_y	1.0	1.1	1.2	1.3	1.4	1.5	1.75	2.0	2.5	3.0
α_x	0.062	0.074	0.084	0.093	0.099	0.104	0.113	0.118	0.122	0.124
α_y	0.062	0.061	0.059	0.055	0.051	0.046	0.037	0.029	0.020	0.014

(a) *Slabs simply supported on four sides.* For a simply-supported slab, where adequate provision is not made to resist torsion at the corner and to prevent the corners from lifting, the bending moments at mid-span should be assumed to have the values given by the following equations:

$$M_x = \alpha_x w l_x^2 \tag{4.1}$$

$$M_y = \alpha_y w l_x^2 \tag{4.2}$$

in which M_x and M_y are the bending moments at mid-span on strips of unit width and span l_x and l_y respectively; w is the total load per unit area; l_y is the length of the longer side; l_x is the length of the shorter side; and α_x and α_y are coefficients, derived from Table 4.3 (see also Table 8 in NCP 1, Part 1, 1973).

(b) *Slabs restrained on four sides.* In this case it is assumed that the corners of the slab are prevented from lifting and that adequate provision has been made for torsion. The slab is imagined as consisting of strips — the middle and edge strips as shown in Fig. 4.1. The width of the middle strip is three-quarters of the entire width of the slab, and the edge strip has a width which is one-eighth of the width of the slab. There is an exception: when

Fig. 4.1 Division of slab into middle and edge strips

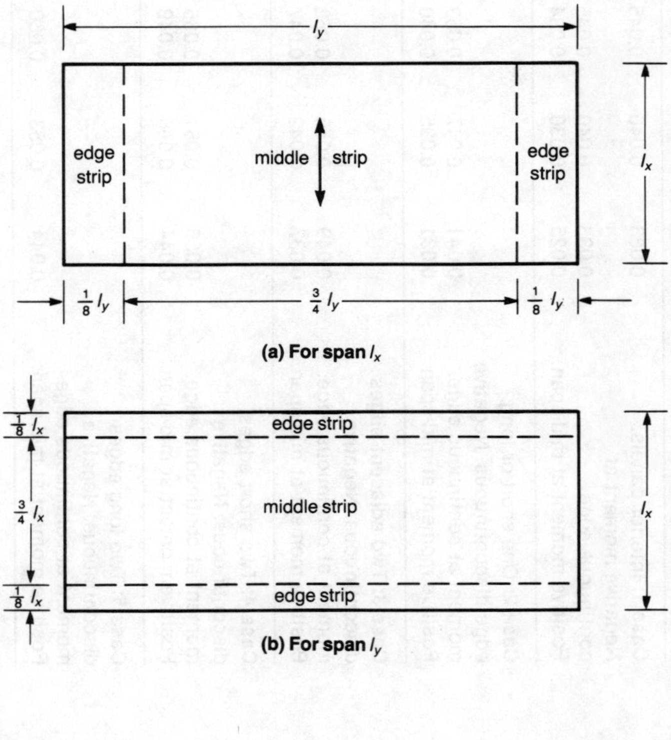

Table 4.4 Bending moment coefficients for rectangular panels supported on four sides, with provision for torsion at the corners

Type of panel and moments considered	Short span coefficient β_x								Long span coefficients β_y for all values of l_y / l_x
	Values of l_y/l_x								
	1.0	1.1	1.2	1.3	1.4	1.5	1.75	2.0 or more	
Case 1: Interior panels.	0.033	0.040	0.045	0.050	0.054	0.059	0.071	0.083	0.033
Negative moment at continuous edge.	0.033	0.040	0.045	0.050	0.054	0.059	0.071	0.083	0.033
Positive moment at mid-span.	0.025	0.030	0.034	0.038	0.041	0.045	0.053	0.062	0.025
Case 2: One short or long edge discontinuous. Negative moment at continuous edge.	0.041	0.047	0.053	0.057	0.061	0.065	0.075	0.085	0.041
Positive moment at mid-span.	0.031	0.035	0.040	0.046	0.046	0.049	0.056	0.064	0.031
Case 3: Two adjacent edges discontinuous. Negative moment at continuous edge.	0.049	0.056	0.062	0.066	0.070	0.073	0.082	0.090	0.049
Positive moment at mid-span.	0.037	0.042	0.047	0.050	0.053	0.055	0.062	0.068	0.037
Case 4: Two short edges discontinuous. Negative moment at continuous edge.	0.056	0.061	0.065	0.069	0.071	0.073	0.077	0.080	
Positive moment at mid-span.	0.044	0.046	0.049	0.051	0.053	0.055	0.058	0.060	0.044
Case 5: Two long edges discontinuous. Negative moment at continuous edge.									0.056
Positive moment at mid-span.	0.044	0.053	0.060	0.065	0.068	0.071	0.077	0.080	0.044
Case 6: Three edges discontinuous (one short or long edge continuous). Negative moment at continuous edge.	0.058	0.065	0.071	0.077	0.081	0.085	0.092	0.098	0.058
Positive moment at mid-span.	0.044	0.049	0.054	0.058	0.061	0.064	0.069	0.074	0.044
Case 7: Four edges discontinuous. Positive moment at mid-span.	0.050	0.057	0.062	0.067	0.071	0.075	0.081	0.083	0.050

For values of l_y/l_x other than those given in Table 4.4 the coefficient β_x and β_y can be obtained by a simple linear interpolation.

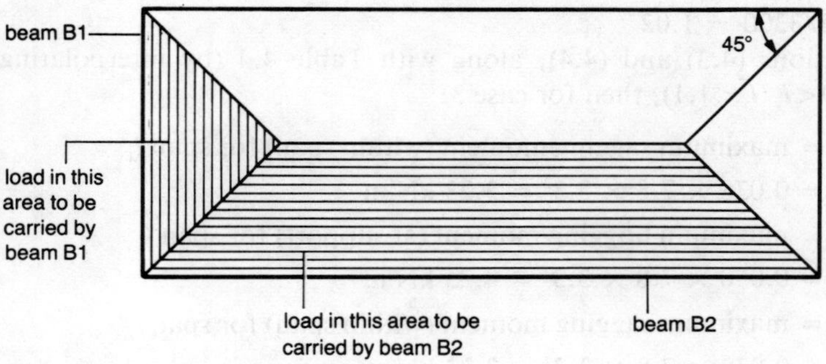

Fig. 4.2 Division of slab load for supporting beams

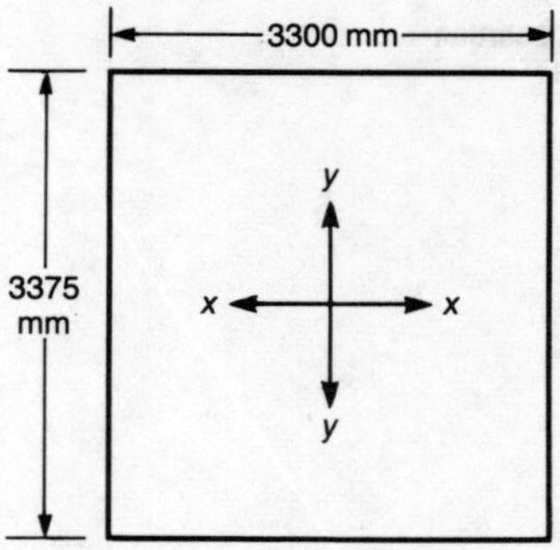

Fig. 4.3 Floor slab for residential apartment (Example 4.2)

$l_y/l_x > 4.0$ then the middle strip in the short direction should be assumed to have a width of $l_y - l_x$ and each edge strip should be assumed to have a width of $l_x/2$.

The maximum bending moments per unit width in the middle strip of a slab are given by:

$$M_x = \beta_x w l_x^2 \tag{4.3}$$

and

$$M_y = \beta_y w l_x^2 \tag{4.4}$$

where M_x and M_y are the maximum bending moments on strips of unit width in the direction of spans l_x and l_y respectively (see Fig. 4.1); w is the total load per unit area; and β_x and β_y are coefficients, some of which are shown in Table 4.4 (from Table 9, NCP 1, 1973).

Where the end of a slab is connected monolithically to a beam or a wall, provision must be made for the negative moments that may occur in the slab at such a support. The negative moment to be assumed will depend on the degree of fixity at the connection. The NCP 1 recommendations are, for general purposes, that this negative moment be taken as two-thirds of the moment given by equations (4.3) and (4.4) for the middle span of the slab. (c) The third recommendation for method 2 is that the loads on the supporting beams are assumed to be in accordance with the distribution shown in Fig. 4.2.

Method 3: This is based on the load-factor method which is treated in detail in chapter 7.

Example 4.2

One of the floor slabs for a residential apartment is rectangular and has dimensions 3300 mm × 3375 mm (see Fig. 4.3). The slab is continuous over two beams that are adjacent to one other. The other two sides are discontinuous. The architect has recommended a slab thickness of 150mm. Check whether the stipulated thickness is adequate and proceed to determine the amount of reinforcement required. Assume $p_{st} = 126\,\text{N/mm}^2$, $p_c = 7\,\text{N/mm}^2$ and $m = 15$. The load intensity on the slab has been estimated to be $7.8\,\text{kN/m}^2$.

Solution

$l_y/l_x = 3375/3300 = 1.02$

Using equations (4.3) and (4.4), along with Table 4.4 (by interpolating between $1.0 < l_y/l_x < 1.1$), then for case 3:

$$M_x^+ = \text{maximum sagging moment (within span) for span } l_x$$
$$= 0.038 \times 7.8 \times 3.3^2 = 3.23 \text{ kN m}$$
$$M_x^- = \text{maximum hogging moment (at support) for span } l_x$$
$$= 0.050 \times 7.8 \times 3.3^2 = 4.25 \text{ kN m}$$
$$M_y^+ = \text{maximum sagging moment (within span) for span } l_y$$
$$= 0.038 \times 7.8 \times 3.3^2 = 3.23 \text{ kN m}$$
$$M_y^- = \text{maximum hogging moment (at support) for span } l_y$$
$$= 0.050 \times 7.8 \times 3.3^2 = 4.25 \text{ kN m}$$

With $p_{st} = 126 \text{ N/mm}^2$, $p_c = 7 \text{ N/mm}^2$ and $m = 15$, the moment of resistance of a 150 mm slab (putting $d = 130$ mm, $b = 1000$ mm and $K = 1.34 \text{ N/mm}^2$) is:

$$M_R = 1.34 \times 1000 \times 130^2/10^6 = 22.65 \text{ kN m}$$

But the maximum bending moment on the slab = 4.25 kN m. Therefore a 150 mm slab is acceptable.

$$A_{st} \text{ for } M_x^+ = \frac{3.23 \times 10^6}{126 \times 0.85 \times 130} = 232 \text{ mm}^2$$

From Table 4.2 use 8 mm ϕ bars @ 200 mm c/c (251 mm²) (bottom within span).

$$A_{st} \text{ for } M_x^- \text{ (by proportion)} = 4.25 \times 232/3.23$$
$$= 305 \text{ mm}^2$$

Use 8 mm ϕ bars @ 150 mm c/c (335 mm²) (top at support).

A_{st} for M_y^+ = 8 mm ϕ bars @ 200 mm c/c (bottom within span).

A_{st} for M_y^- = 8 mm ϕ bars @ 150 mm c/c (top at support).

4.5 Slabs continuous over supports

The design of a slab that is continuous over supports is usually based on a unit width strip which runs perpendicularly to the supports (Fig. 4.4).

Normally continuous slabs (and beams) may be considered as capable of free rotations about their supports. However, in situations where the supports are monolithic with (and stiff in relation to) the slab, the design must take this fixity into account.

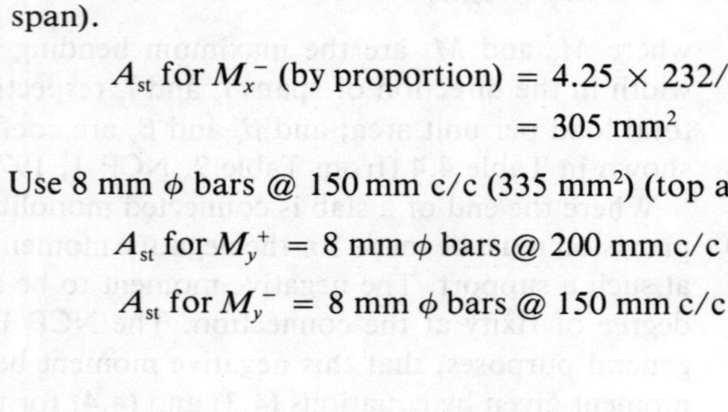

Fig. 4.4 Unit width strip for a slab continuous over supports

Table 4.5 Approximate values of bending moment in uniformly-loaded beams and slabs continuous over three or more approximately equal spans

	Near middle of end span	At support next to end support	At middle of interior end	At other interior support
Moment due to dead load	$W_d l / 12$	$-W_d l / 10$	$W_d l / 24$	$-W_d l / 12$
Moment due to super-imposed load	$W_s l / 10$	$-W_s l / 9$	$W_s l / 12$	$-W_s l / 9$

In the absence of more exact estimates, the bending moments in uniformly-loaded slabs (and beams) that are continuous over three or more approximately equal spans may be assumed to have the values given in Table 4.5 (quoted from Table 7, NCP 1) in which W_d = total dead load per span and W_s = total superimposed load per span. Two spans may be considered as approximately equal when they do not differ by more than 15 per cent of the longer span.

4.6 Ribbed and hollow slabs

Ribbed and hollow block slabs have been introduced earlier in section 1.2.1. These slabs consist of a series of reinforced concrete ribs cast in situ (that is, on the construction site) between permanent or temporary blocks which may or may not be introduced after the concrete has set. The requirements of NCP 1 are that the blocks which remain part of the slab and are considered to contribute to the structural strength should be either of concrete or burnt clay and should have a crushing strength of at least 17.5 N/mm^2. (The crushing strength of the block is assumed to be measured on the net section when axially loaded in a direction corresponding to that in which they will function in the floor slab.)

A topping of concrete cast in situ over the blocks is usually used for hollow block constructions. The requirements of NCP 1 are that the thickness of the topping should be governed by the following, under the stipulated conditions:

- in floors with permanent blocks regarded as contributing to the strength of the construction (and with a clear distance between the ribs not exceeding 450 mm) minimum thickness = 30 mm. When the blocks are properly jointed (with 1:3 cement-sand mortar), the minimum thickness may be reduced to 25 mm.
- in floors with permanent blocks *not* regarded as contributing to the strength of the construction the minimum thickness = 40 mm or one-twelfth of the clear distance between the ribs, whichever is the greater.
- in all other cases the minimum thickness should be 50 mm.

The minimum width of the rib is 65 mm and the depth, excluding any topping, is to be limited to four times the width. The spacing of the rib is also to exceed 100 mm, centre-to-centre.

In computing the moment of resistance of a hollow block slab the contribution due to the block may be neglected. On the other hand, the blocks may be considered as acting in structural combination with the ribs and topping (when used). The only provision is that the blocks are properly jointed and the thickness of the topping is at least 30 mm. The elastic modulus of the material forming the block may be assumed to be the same as for concrete (NCP 1, clause 4.1.9.4).

With regard to the reinforcement in the ribs the NCP 1 requirements can be summarised thus:

- not less than 50 per cent of the total main reinforcement should be carried through at the bottom onto the bearing and effectively anchored.
- slabs continuous over supports may be treated as simply supported due to the difficulty of providing sufficient reinforcement to carry the full support moment on the basis of continuity.
- when a continuous slab is treated as simply supported, adequate provision must be made to prevent cracking by introducing reinforcement over the supports. The cross-sectional area of such reinforcement must be at least equal to 25 per cent of that in the middle of the adjoining bays.
- the lateral spacing of bars running parallel in the concrete ribs may be reduced to 12 mm or the diameter of the bar, whichever is greater (provided that both the permissible bond and compressive stresses in the rib are reduced by 40 per cent).

With these considerations in mind, the analysis/design of a ribbed or hollow block slab is in line with the procedure outlined earlier for beams and solid slabs.

4.7 Flat slabs

As defined in section 1.2.1 the term flat slab is used for a reinforced concrete slab (with or without drops) supported directly on columns, generally without beams.

There are two major approaches to the design of a flat slab. The first is based on the so-called 'continuous frame' method and the second is an empirical method. A third general-purpose method is based on the finite element method that has been mentioned earlier. In this case, the flat slab is subdivided into meshes or elements (triangular, rectangular or polygonal) and the partial differential equations governing the structural behaviour of plates is solved via this discretisation. A detailed description of the finite element method is obviously beyond the scope of this book but those interested can refer to Gallagher (1975), Zienkiewicz (1977), and Desai and Abel (1972), for a comprehensive treatment of the method.

Before the two common approaches can be considered in detail, it is expedient to subdivide any given flat slab into some standard panels or strips (see Fig. 4.5). The panels consist of the column strips and the middle strips (*cf.* the case of slabs spanning in two-directions as subdivided in Fig. 4.1). The width of the column strip is taken as one half of the width of the entire panel. One exception is where drops are used, in which case the width of the column strip is taken to be equal to that of the drop.

The width of the middle strip is also taken as one half of the width of the entire panel. However, in situations where the column strip is taken as the width of drop, the middle strip is all that remains of the panel width minus the width of the drop.

If L_1 = length of the panel in the direction of the span, L_2 = width of the panel at right angles (measured in each case from the centres of the

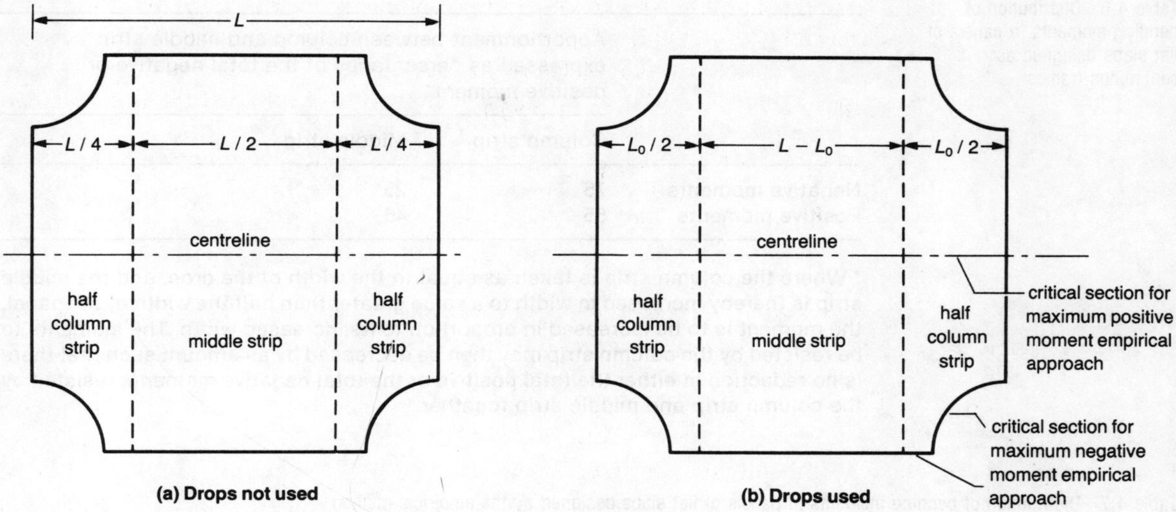

Fig. 4.5 Division of flat slab into column and middle strips

columns) and $L = \frac{1}{2}(L_1 + L_2)$, then NCP 1 requires that the total thickness of a flat slab should not be less than the greatest of the following values:

- 125 mm;
- $L/32$ for end panels without drops;
- $L/36$ for interior panels that are fully continuous without drops, and for end panels with drops;
- $L/40$ for interior panels that are fully continuous, with drops.

4.7.1 The continuous frame method

In this method the flat slabs are considered as continuous frames. The bending moments and shearing forces are determined on the basis of the following assumptions (NCP 1, clause 4.3.10.2):

- the structure may be considered to be divided longitudinally and transversely into frames consisting of a row of columns and strips of slab, with a width equal to the distance between the centrelines of the panels on each side of the row of columns.
- each frame may be analysed by the Hardy Cross (or other suitable) method, in its entirety; or each strip of slab may be analysed as a separate frame, with the columns above and below assumed to be fixed at their extremities. The spans used in the analyses should be the distance between the centres of the supports, except where the slab is supported by a wall when the span should be the distance of the face of the wall plus one-half of the depth of the slab.

The requirements of NCP 1 are that the sum of the maximum positive bending moment and the average of the negative bending moments, used in the design of any one span of the slab should, for the whole panel width, be not less than $wL_2(L_1 - 2D/3)^2/10$ where w is the total load per unit area on the panel and D is the diameter of the column heads supporting the slab. [Clause 4.3.9 of NCP 1 provides the minimum requirements in the choice of D (eg. $D \not> 0.25L$).] The bending moments for which provision is made should be divided between the column and the middle strips in the proportions given in Table 4.6 (quoted from Table 13 of NCP 1, 1972).

Table 4.6 Distribution of bending moments in panels of flat slabs designed as continuous frames

	Apportionment between column and middle strip expressed as percentages of the total negative or positive moment*	
	Column strip	Middle strip
Negative moments	75	25
Positive moments	55	45

* Where the column strip is taken as equal to the width of the drop, and the middle strip is thereby increased in width to a value greater than half the width of the panel, the moment is to be increased in proportion to its increased width. The moments to be resisted by the column strip may then be decreased by an amount such that there is no reduction in either the total positive or the total negative moments resisted by the column strip and middle strip together.

Table 4.7 Distribution of bending moments in panels of flat slabs designed by the empirical method

		Apportionment of moments between the column and middle strips expressed as a percentage of M_0	
		Column strip	Middle strip
Interior panels with drops	Negative moments	50	15
	Positive moments	20	15
Interior panels without drops	Negative moments	46	16
	Positive moments	22	16
Exterior panels with drops	Exterior negative moments	45	10
	Positive moments	25	19
	Interior negative moments	50	15
Exterior panels without drops	Exterior negative moments	41	10
	Positive moments	28	20
	Interior negative moments	46	16

4.7.2 The empirical method

Where the empirical approach is followed, the code places certain limitations on the numbers/shapes of the series of panels used and the shape/sizes of the drops. These limitations are (see NCP 1, clause 4.3.11.2):

- the slabs should comprise a series of rectangular panels of approximately constant thickness, arranged in at least three rows in two directions at right-angles, and the ratio of the length of a panel to its width should not exceed 4/3. The length and/or widths of any two adjacent panels in a series should not differ by more than 10 per cent of the greater length or width. End spans may be shorter, but not longer than, interior spans. Where adjacent spans differ, the length should always be taken as that of the longer span when calculating the bending moments.
- the drops should be rectangular in plan and have a length in each direction not less than one third of the panel length in that direction. For exterior panels the width of drop at right angles to the non-continuous

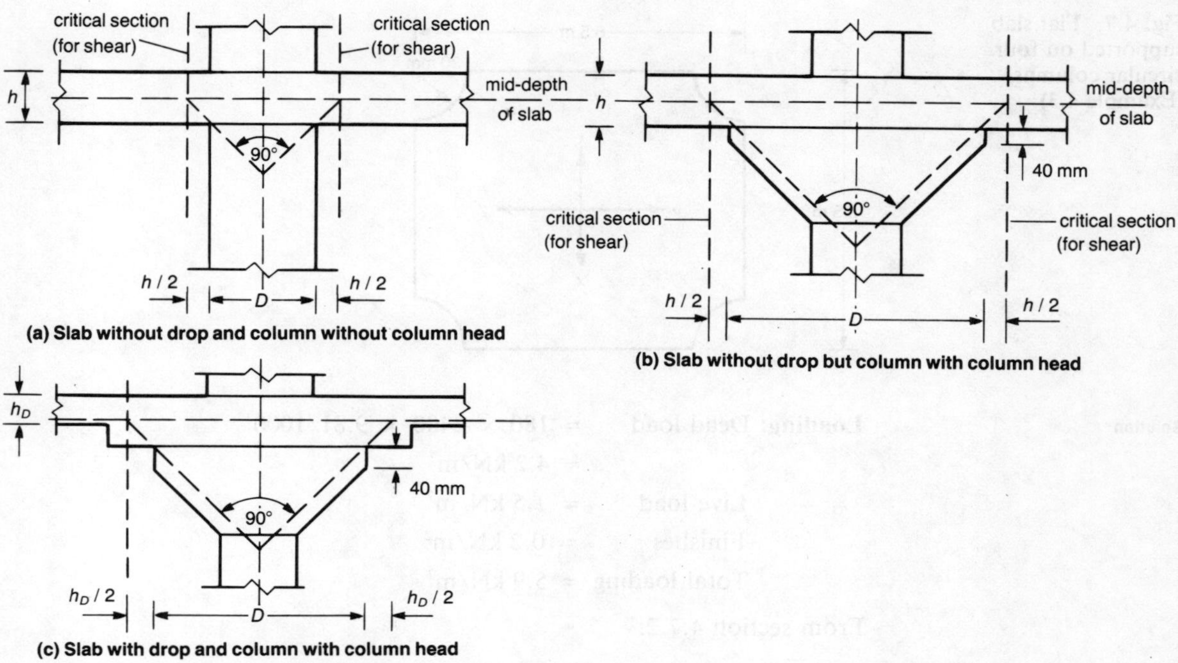

Fig. 4.6 Critical section for shearing stresses in flat slabs

edge and measured from the centreline of the columns should be equal to one half the width of drop for interior panels.

For interior panels that are fully continuous the critical sections for bending moments are the centrelines of the panel (for positive moments) and along the edges of the panel, on lines joining the centres of the columns and around the perimeter of the column heads. These critical sections are as indicated in Fig. 4.5(b).

The design bending moments are apportioned to the column and middle strips as percentages of the moment $M_0 = wL_2(L_1 - 2D/3)^2/10$. The distribution of bending moments is presented in Table 4.7 (quoted from Table 14, NCP 1). In using Table 4.7 it must be noted that for flat slabs with drops, where the column strip is taken as being equal to the width of the drop [Fig. 4.5(b)], the moments in the middle strip (if the width is now greater than half the entire panel width) must be increased in proportion to the increased width.

4.7.3 Shear stresses

The critical sections for shear are shown in Fig. 4.6. The dimensions of the slab, column head and drop (where used) must be chosen such that the permissible shear stresses are not exceeded (see Table 2.1 for values for nominal concrete mixes).

Example 4.3

Using the empirical method obtain the maximum bending moments developed in a flat slab (5.5 m × 6.5 m × 180 mm) that is supported directly on four circular columns each 60 mm in diameter (Fig. 4.7). The slab is the floor of an interior bedroom in a residential building and the finishes have been estimated to weigh 0.2 kN/m^2.

Fig. 4.7 Flat slab supported on four circular columns (Example 4.3)

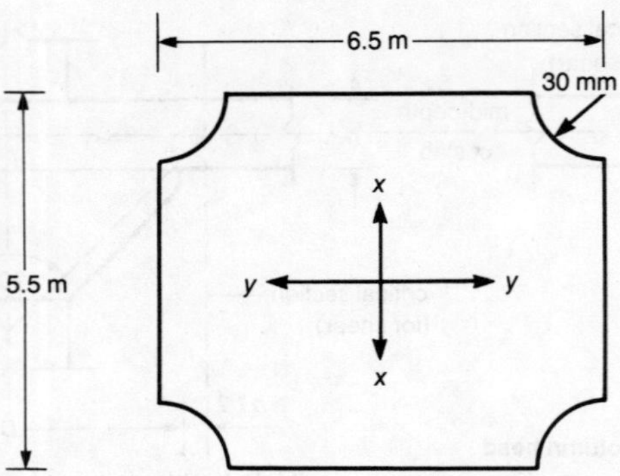

Solution

Loading: Dead load $= 180 \times 2400 \times 9.81/1000^2$
$= 4.2\ \text{kN/m}^2$

Live load $= 1.5\ \text{kN/m}^2$

Finishes $= \underline{0.2}\ \text{kN/m}^2$

Total loading $= \underline{5.9}\ \text{kN/m}^2$

From section 4.7.2:

$$M_0 = wL_2(L_1 - 2D/3)^2/10.$$

Therefore, for this problem $w = 5.9\ \text{kN/m}^2$.

For the 5.5 m span, $L_1 = 5.5$ m, $L_2 = 6.5$ m, D = 60 mm and:

$$M_0 = M_{0x} = 5.9 \times 5.6\,(5.5 - 2 \times 60 \times 10^{-3}/3)^2/10$$
$$= 114.3\ \text{kN m}$$

For the 6.5 m span, $L_1 = 6.5$ m, $L_2 = 5.5$ m and

$$M_0 = M_{0y} = 5.9 \times 5.5\,(6.5 - 2 \times 60 \times 10^{-3}/3)^2/10$$
$$= 135.4\ \text{kN m}$$

Using Table 4.7 (see portion referring to interior panels without drops) it can be seen that, for the 5.5 m span:

M_c^- = maximum negative moment in column strip
$= 0.46 \times M_{0x} = 52.6$ kN m

M_c^+ = maximum positive moment in column strip
$= 0.22 \times M_{0x} = 25.1$ kN m

M_m^- = maximum negative moment in middle strip
$= 0.16 \times M_{0x} = 18.3$ kN m

M_m^+ = maximum positive moment in middle strip
$= 0.16 \times M_{0x} = 18.3$ kN m

and for the 6.5 m span:

M_c^- = maximum negative moment in column strip
$= 0.46 \times M_{0y} = 62.3$ kN m

M_c^+ = maximum positive moment in column strip
$= 0.22 \times M_{0y} = 29.8$ kN m

$M_m^- =$ maximum negative moment in middle strip

$= 0.16 \times M_{0y} = 21.7$ kN m

$M_m^+ =$ maximum positive moment in middle strip

$= 0.16 \times M_{0y} = 21.7$ kN m

Comments

Now that the moments have been distributed to the various strips, the widths of which are known (see Fig. 4.5), the appropriate reinforcement can be obtained in the same way as for beams and slabs. The next important considerations are shear and bond, which are dealt with in sections 4.10 and 6.1, respectively.

4.8 Staircases

The various parts of a typical staircase are shown in Fig. 4.8. These include the tread T, riser R, nosing N, and waist W.

In respect of loadings on stairs the requirements of NCP 1 (clause 4.6.1) are that:

- for stairs with open wells (see Fig. 4.9) where spans partly cross at right angles (eg, the lower span in Fig. 4.8 partly crosses the open span at right angles at the half landing), the load on areas common to any two of such spans may be taken as one half in each direction, as shown in Fig. 4.9 (w = intensity of loading).

Fig. 4.8 Typical staircase

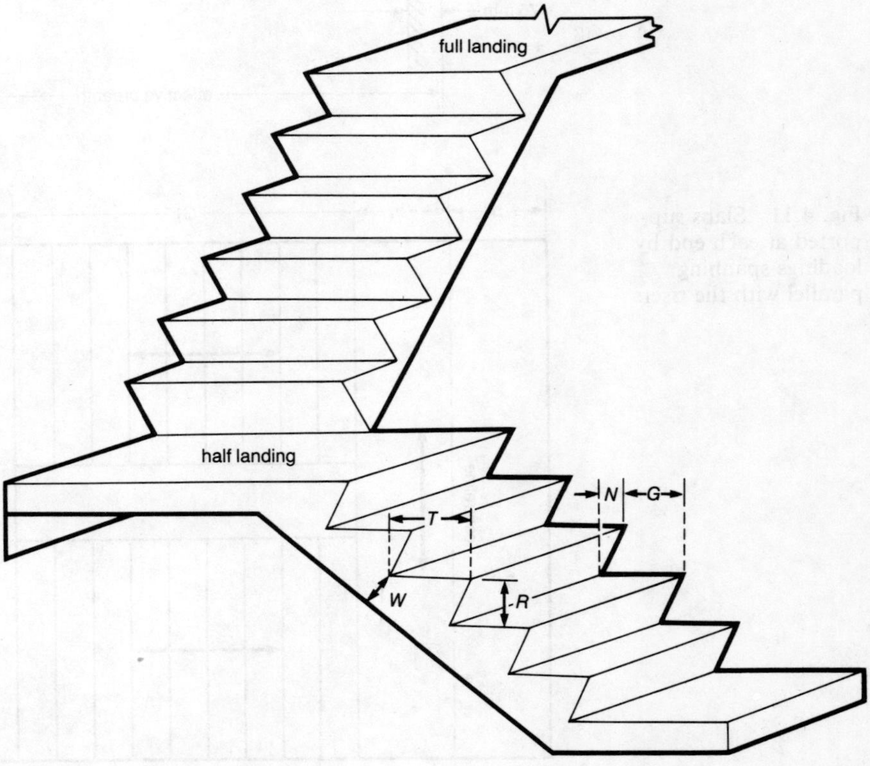

Fig. 4.9 Loading on stairs with open wells

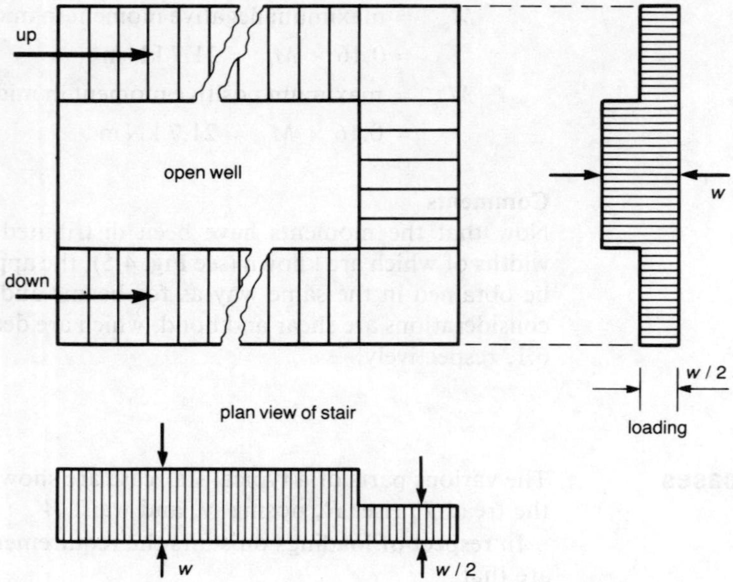

Fig. 4.10 Loading on stairs built into walls

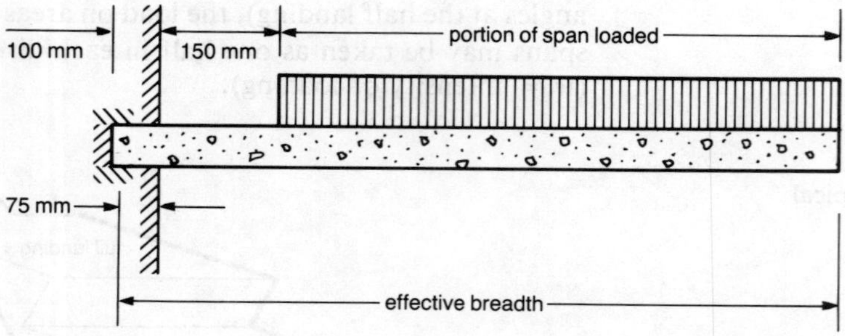

Fig. 4.11 Slabs supported at each end by loadings spanning parallel with the risers

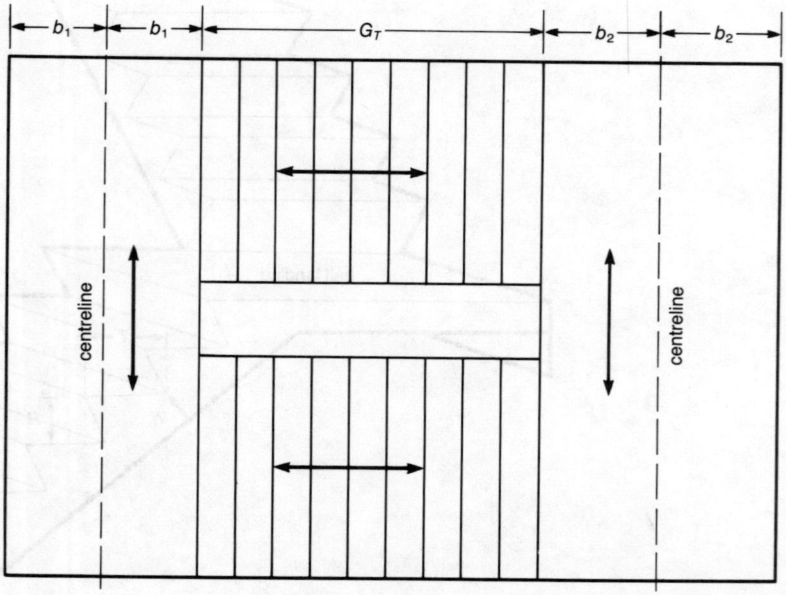

- where flights or landings are built a distance of not less than 100 mm into walls and are designed to span in the direction of the flight, a 150 mm strip may be deducted from the loaded area and the effective breadth of the section increased by 75 mm, as depicted in Fig. 4.10.

Clause 4.6.2 (NCP 1, 1972) refers to the effective span of stairs without stringer beams (ie, beams running along the underside of the flight of stairs). The effective spans of such stairs are to be taken as the following distances:

- where supported at top and bottom risers by beams spanning parallel with the risers, a distance equal to the total 'going' G_T of the stairs plus, at each end, either half the width of the landing or 1.0 m, whichever is the smaller. This recommendation is summarised in Table 4.8 for all combinations of half-landing widths b_1 and b_2 (see also Fig. 4.11).

Table 4.8 Effective spans for various combinations of half landing widths

	b_1 (m)	b_2 (m)	Effective span (m)
Case 1	< 1.0	< 1.0	$G_T + b_1 + b_2$
Case 2	< 1.0	≥ 1.0	$G_T + b_1 + 1.0$
Case 3	≥ 1.0	< 1.0	$G_T + b_2 + 1.0$
Case 4	≤ 1.0	≥ 1.0	$G_T + 2.0$

- where the landing slabs span in the same direction as the stairs, they should be considered as acting together to form a single slab and the span determined as indicated in the preceding paragraph.

The design procedure for a staircase is along the same line as established earlier for regular solid slabs, the main difference being in the computation of loads and the choice of the effective spans. Some of the important considerations in the design of staircases are highlighted in the following examples.

Example 4.4

The staircase in a two-storey building is to consist of 300 mm treads, 150 mm risers and 275 mm goings. The staircase is to span a horizontal distance of 2.4 m while resting on beams at the top and bottom of the flight. If the total rise of the stairs is 2.25 m and the finishes are estimated to weigh 0.5 kN/m², obtain a suitable waist thickness and the required amount of reinforcement. Assume $p_{st} = 126\,\text{N/mm}^2$, $p_c = 7\,\text{N/mm}^2$ and $m = 15$.

Solution

The loading is to be expressed as the resultant of the sum of all the loads acting on the staircase (which is inclined), measured in plan. To this end a load conversion factor β is introduced such that (see Fig. 4.12) the load in plan = the load on the inclined staircase multiplied by the load conversion factor, or $w = \beta w'$.

The obvious requirements are that:

$$w'l' = wl$$

so that:

$$w = w'l'/l = w' \sec\theta$$

Fig. 4.12 Load conversion from 'inclined plane' to 'plan'

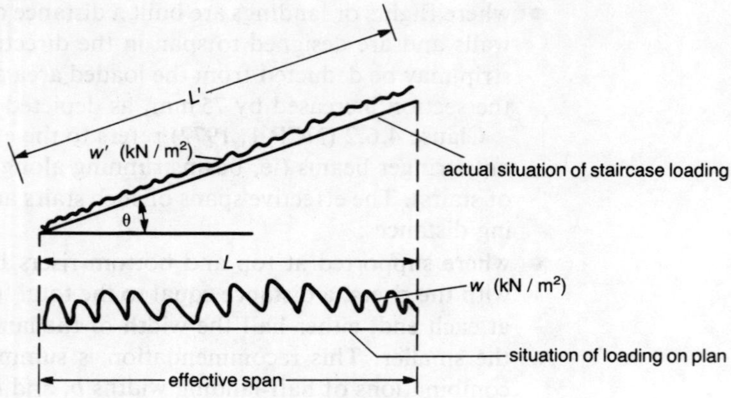

That is, $\beta = \sec\theta$, where θ is the inclination of the staircase to the horizontal. In terms of the riser R, and the going G:

$$\beta = \sqrt{(R^2 + G^2)}/G$$

For this particular problem $R = 0.15$ m and $G = 0.275$ m so:

$$\beta = \sqrt{(0.15^2 + 0.275^2)}/0.275 = 1.14$$

Loading:

Dead load due to steps (in plan) $= \frac{1}{2} \times 0.15 \times 23.6$
$= 1.77\ \text{kN/m}^2$

Dead load due to waist (assume 100 mm thickness) $= 1.14 \times 0.1 \times 23.6$
$= 2.69\ \text{kN/m}^2$

Dead load due to finishes $= 0.5 \times 1.14 = 0.57\ \text{kN/m}^2$

Therefore, total dead load $= 1.77 + 2.69 + 0.57 = 5.03\ \text{kN/m}^2$

Live load (see Table 4.1) $= 1.5\ \text{kN/m}^2$ in plan

Considering a unit width of staircase and using Table 4.5 with $W_d = 5.03 \times 2.4 \times 1 = 12.07$ kN and $W_s = 1.5 \times 2.4 \times 1 = 3.6$ kN, then:

maximum negative moment at support $= 12.07 \times 2.4/10 + 3.6 \times 2.4/9$
$= 3.86$ kN m

maximum positive moment within span $= 12.07 \times 2.4/12 + 3.6 \times 2.4/10$
$= 3.28$ kN m

Assuming a balanced section with $n = 0.45$, $a = 0.85$ and $K_c = 1.34\ \text{N/mm}^2$, then the required value of d is given by:

$$d = \sqrt{\frac{3.86 \times 10^6}{1.34 \times 10^3}} = 53.7\ \text{mm}$$

Allowing for 21 mm cover to the centre of the reinforcement, the minimum value of $h = 54.6 + 21 = 74.7$ mm. Therefore, the waist should be taken as 80 mm.

The value of A_{st} required for the supports is given by:

$$A_{st} = \frac{3.86 \times 10^6}{126 \times 0.85 \times (80 - 21)} = 611 \text{ mm}^2$$

From Table 4.2, use 12 mm ϕ bars @ 175 mm c/c (646 mm^2).

The value of A_{st} required for distribution is given by:

$$A_{st} = \tfrac{1}{4} \times 611 = 153 \text{ mm}^2$$

or 6 mm ϕ bars @ 175 mm c/c (162 mm^2).

The value of A_{st} required within the span is given by:

$$A_{st} = \frac{3.4}{3.99} \times 611 = 519 \text{ mm}^2$$

Use 12 mm ϕ bars @ 200 mm c/c (565 mm^2).

The value of A_{st} required for distribution is given by:

$$A_{st} = \tfrac{1}{4} \times 519 = 130 \text{ mm}^2$$

or 6 mm ϕ bars @ 200 mm c/c (141 mm^2).

Example 4.5

The flight of stairs in a busy Federal Ministry is shown in plan in Fig. 4.13. It consists of 275 mm treads, 160 mm risers and 250 mm goings. If the finishes weigh 0.35 kN/m^2, carry out the design of the staircase. Assume $p_{st} = 140$ N/mm^2, $p_c = 8.5$ N/mm^2 and $m = 15$.

Solution

In this problem $b_1 = 1.2/2 = 0.6 \text{ m} < 1.0 \text{ m}$ and $b_2 = 2.4/2 = 1.2 \text{ m} > 1.0 \text{ m}$. Therefore the effective span must be selected according to the requirement for case 2 in Table 4.8, that is:

Effective span $l = G_T + b_1 + 1.0$

or, with $G_T = 2.5$ m

$l = 2.5 + 0.6 + 1.0 = 4.1$ m

Now, with $R = 160$ mm, $G = 250$ mm and $T = 275$ mm, the load conversion factor β is given by:

$$\beta = \sqrt{(R^2 + G^2)}/G$$
$$= \sqrt{(0.16^2 + 0.25^2)}/0.25$$
$$= 1.19$$

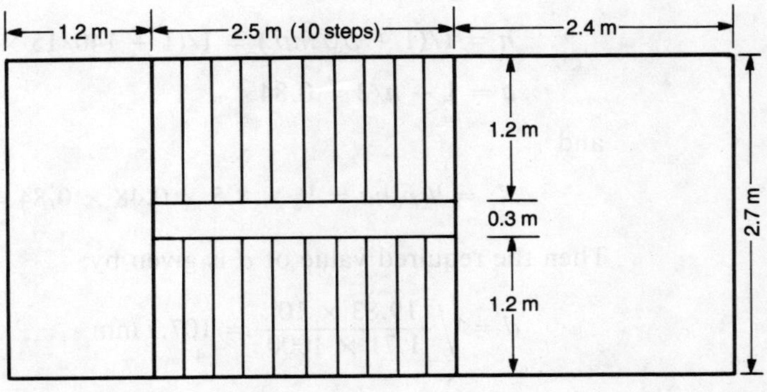

Fig. 4.13 Flight of stairs in busy Federal Ministry (Example 4.5)

Loading:

Dead load due to steps $= \frac{1}{2} \times 0.16 \times 23.6$
$= 1.89\ kN/m^2$

Dead load due to waist (assuming 140 mm thickness) $= 1.19 \times 0.14 \times 23.6$
$= 3.93\ kN/m^2$

Dead load due to finishes $= 0.35 \times 1.19 = 0.42\ kN/m^2$

Therefore, total dead load $= 1.89 + 3.93 + 0.42 = 6.24\ kN/m^2$

Live load (assuming that the staircase will be heavily used) $= 5.0\ kN/m^2$

Therefore, total load on the stairs $= 6.24 + 5.0 = 11.24\ kN/m^2$

The loaded unit strip of the staircase is shown in Fig. 4.14. Note that the portions of the span contributed by the landings are unloaded because the landings are spanned in a different direction (see Fig. 4.10).

Taking moments about support B:

$$4.1 \times R_A = 11.24 \times 2.5\,(2.5/2 + 1.0)$$

or

$$R_A = 15.42\ kN$$

and

$$R_B = 11.24 \times 2.5 - 15.42$$
$$= 12.68\ kN$$

At any distance $y \geq 0.6$ m from support A the bending moment is:

$$M_y = 15.42y - 11.24(y - 0.6)^2/2$$

which is maximum when:

$$15.42 = 11.24(y - 0.6)$$

or

$$y = 1.97\ m$$

Therefore, $M_{max} = 15.42 \times 1.97 - 11.24(1.97 - 0.6)^2/2$
$= 19.83\ kN\ m$

Assuming a balanced section with:

$$n = 1/(1 + p_{st}/mp_c) = 1/(1 + 140/15 \times 8.5) = 0.48$$

$$a = 1 - n/3 = 0.84$$

and

$$K_c = \tfrac{1}{2} p_c n a = \tfrac{1}{2} \times 8.5 \times 0.48 \times 0.84 = 1.71\ N/mm^2$$

Then the required value of d is given by:

$$d = \sqrt{\frac{19.83 \times 10^6}{1.71 \times 1000}} = 107.7\ mm$$

Fig. 4.14 Loaded unit strip of staircase (solution to Example 4.5)

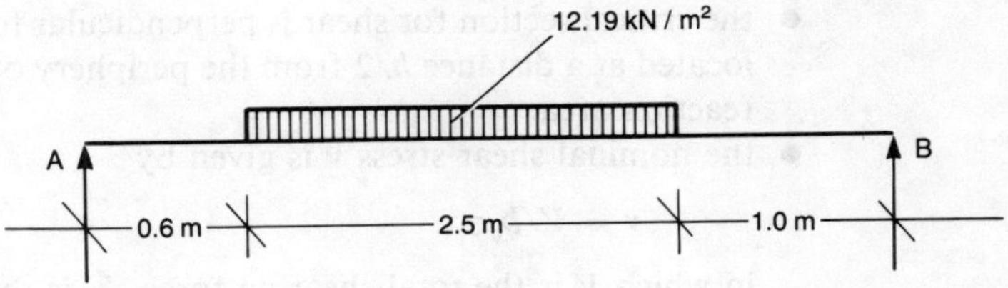

With a 24 mm total cover to the centre of the reinforcement

$$h = 107.7 + 24 = 131.7 \text{ mm}$$

Therefore, a waist 140 mm thick should be provided.

The value of A_{st} required is given by:

$$A_{st} = \frac{19.83 \times 10^6}{140 \times 0.84 \times (140 - 24)} = 1454 \text{ mm}^2$$

Use 16 mm ϕ bars @ 125 mm c/c (1608 mm²). For distribution use 10 mm ϕ bars @ 200 mm c/c (393 mm²). (Refer to section 13.4 for the structural detailing of staircases).

4.9 Deflection

Maximum span/depth ratios to prevent excessive deflections of slabs are stipulated in NCP 1 and these are reproduced in Table 4.9. If the span/depth ratio is computed from bending moment considerations and is found to exceed the values in Table 4.9 then deflection is used as the basis for selecting the slab thickness. In such cases, the thickness is chosen such that the span/depth ratio is, at most, equal to the ratios in Table 4.9.

Table 4.9 Maximum span/depth ratios for slabs

Solid one-way slabs	Support conditions			
	simply supported	one end continuous	both ends continuous	cantilever
	25	30	35	12

4.10 Shear

The requirements of NCP 1 (clause 3.13.1) are that the shear capacity of slabs (and footings) in the vicinity of concentrated loads or concentrated reactions be governed by the more severe of the following two conditions:

- the slab (or footing) acting essentially as a wide beam, with a potential diagonal crack extending in a plane across the entire width. (This case is similar to that outlined for beams in section 3.3 of this book.)
- two-way action existing for the slab (or footing), with potential diagonal cracking along the surface of a truncated cone or pyramid around the concentrated load or reaction.

In the latter case the requirements for shear design are outlined overleaf:

- the critical section for shear is perpendicular to the plane of the slab and located at a distance $h/2$ from the periphery of the concentrated load or reaction area.
- the nominal shear stress v is given by

$$v = V/b_0 z \tag{4.5}$$

in which V is the total shearing force, b_0 is the periphery of the critical section for slabs (and footings), and z is the lever arm.
- if U = the compressive stress in concrete, then the shear stress v, computed from equation (4.5) must not exceed $U^{1/2}$. Otherwise shear reinforcement must be provided. Under no circumstances must the shear stress v exceed $U^{1/3}$.
- shear reinforcement must be in accordance with the requirements set earlier for beams (see section 3.3), except that the permissible stress in the shear reinforcement is to be one-half of the permissible stress in bending.
- shear reinforcement consisting of bars, rods or wires cannot be considered effective in a slab whose thickness is less than 254 mm.

Example 4.6

Determine the maximum shear developed in the flat slab of Example 4.3, assuming a two-way action exists for the slab. Assume p_{st} = 126 N/mm², p_c = 7 N/mm² and m = 15.

Solution

The critical section for shear is at distance $h/2$ = 90 mm from the face of the column. If it is borne in mind that an interior panel is being dealt with, the diameter of the critical plane is 60 + 90 = 150 mm and the net shear on the section is:

$$V = 5.9 \times (5.5 \times 6.5 - \pi \times 0.15^2/4) = 210.9\,\text{kN}$$

Therefore, the maximum shear stress is:

$$v = \frac{210.8 \times 10^3}{\pi \times 150 \times 0.85 \times (180 - 21)} = 3.3\,\text{N/mm}^2$$

Comments

The maximum shear stress has exceeded the permissible amount ($\sqrt{7}$ = 2.6 N/mm²) and a remedy for the situation may be to increase the slab thickness by providing a drop and/or introducing a flared column head.

Suppose a drop is introduced such that h = 300 mm, then:

$$v = \frac{210.8 \times 10^3}{\pi \times 150 \times 0.85 \times (300 - 21)} = 1.9\,\text{N/mm}^2 < 2.6\,\text{N/mm}^2$$

4.11 Exercises

(Unless otherwise stated assume p_{st} = 126 N/mm², p_c = 7 N/mm² and m = 15).

4.1 A 5 m × 3 m reinforced concrete platform is to be erected by ABC Bookshop in one of their stores in Lagos. It is expected that, at full capacity, books will be stacked up to a height of 4.5 m. The platform will be cast monolithically with four edge-beams, each of width 250 mm. Carry out a detailed design of the platform by:
(a) spanning in one direction;
(b) spanning in two directions.

4.2 A 4 m × 4 m reinforced concrete slab rests along its edges on block-walls (thickness = 225 mm). The slab is to carry a uniformly-distributed load of 12.5 kN/m^2 and a point load of 95 kN at its centre. Determine a suitable slab thickness and the reinforcement required. (**Hint**: The design must incorporate bending moment, deflection and shear considerations.)

4.3 Fig. Q4.3 shows a rectangular slab which runs over four parallel rectangular beams. If the loading has been estimated to be 9.3 kN/m^2, determine the required slab thickness and reinforcement.

Fig. Q4.3

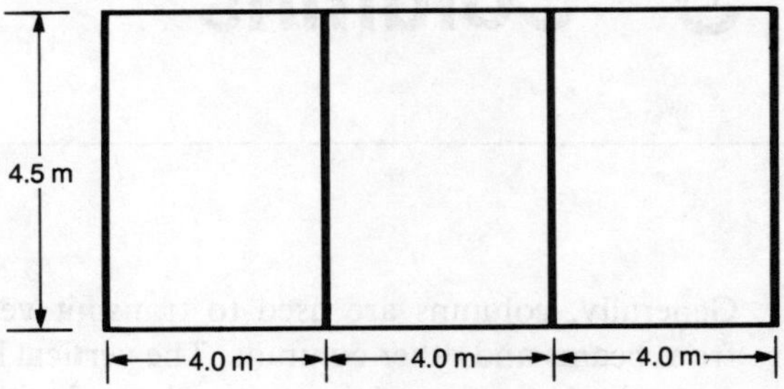

4.4 Obtain the maximum bending moments developed in a flat slab (6.0 m × 6.5 m × 250 mm) supported directly on four circular columns each of diameter 75 mm. The slab is an interior dining hall floor of a five-star hotel.

4.5 What is the maximum shear stress developed in the flat slab of Exercise 4.4? Will the result change if the circular columns are replaced by square ones of side 75 mm?

4.6 The staircase in a three-story office building is to consist of 280 mm treads, 140 mm risers and 250 mm goings. The staircase spans a horizontal distance of 2.5 m and rests on beams at the top and bottom of the flights. Given that the total rise of stairs is 2.24 m determine the required waist thickness and reinforcement.

4.7 The plan view of the flight of stairs in a residential building in Ibadan is shown in Fig. Q4.7. Perform a detailed design of the staircase, assuming it consists of 260 mm treads, 150 mm risers and 220 mm goings. Take the weight of finishes to be 0.65 kN/m^2.

Fig. Q4.7

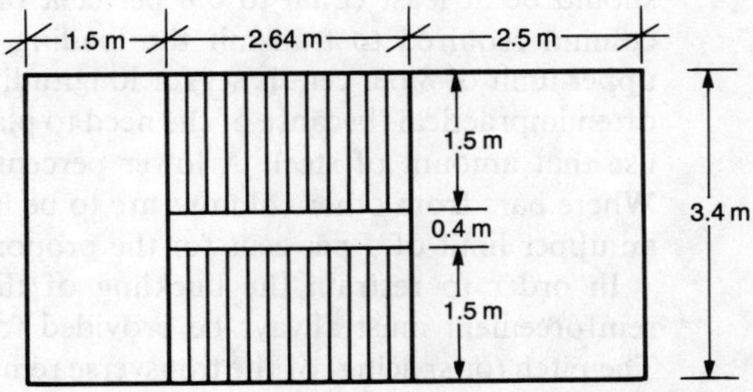

5 Columns

5.1 Introduction

Generally, columns are used to transmit vertical reactions and moments from beams and other columns. The vertical loads may be axial (ie, they act through the main axis, passing through the centroid of the column) or eccentric (ie, the point of action is displaced some distance from the centroidal axis). Any given column transmitting vertical loads and bending moments can be treated as an eccentrically-loaded column, in which case an artificial eccentricity is assumed.

A column can be classified as either short or long. According to NCP 1, a short column is one whose ratio of effective length to least lateral dimension is less than 15. A long column has an effective length that is between 15 and 57 times the least lateral dimension. The classification is important because there is a fundamental difference between the behaviour of a short column and that of a long one when each is loaded.

A long column has a tendency to fail by buckling when loaded. Exact analytical solutions are available to model the buckling behaviour if the column is assumed to be perfectly elastic. Such analyses lead to the so-called Euler's equations, which may be used to simulate the behaviour of reinforced concrete elements subjected to direct compression (eg, columns and beams). It is often necessary (by virtue of the underlying restrictive assumptions) in deriving the Euler's equations to resort to pseudo-exact formulae for the analysis of long columns.

The requirements of NCP 1 (clause 4.2.1) are that a reinforced column should have longitudinal reinforcement with a cross-sectional area that should be at least equal to 0.8 per cent of the cross-sectional area of the column required to transmit the loading (as specified in the Code). An upper limit of 8 per cent is set for longitudinal reinforcement, although it is often impractical (because of the need to place and compact the concrete) to use that amount of steel. A lower percentage is therefore recommended. Where bars from other columns are to be lapped with those in the column, an upper limit of 4 per cent for the proportion of steel is recommended.

In order to restrain the buckling of the longitudinal steel, transverse reinforcement must always be provided for reinforced concrete columns. The pitch (or spacing) of the transverse reinforcement should not exceed the least of the following distances:

- the least lateral dimension of the column;

- twelve times the diameter of the smallest longitudinal reinforcement in the column; or
- 300 mm.

The effective lengths of columns are to be based on the guidelines given in Table 5.1. The length L indicated in the table refers either to the column length measured from floor to floor or the distance between well-restrained supports.

Table 5.1 Effective lengths of columns

Type of column	Effective column length
Properly restrained at both ends, in both position and direction.	$0.75L$
Properly restrained at both ends in position and imperfectly restrained in direction at one or both ends.	A value intermediate between $0.75L$ and L, depending on the efficiency of the directional restraint.
Properly restrained at one end in both position and direction, and imperfectly restrained in both position and direction at the other end.	A value intermediate between L and $2L$, depending on the efficiency of the imperfect restraint.

The Code also requires that the cover to the longitudinal reinforcement in a column be not less than 40 mm nor less than the diameter of the reinforcement bars. Special consideration is given to columns with minimum dimensions of 200 mm or under, whose reinforcement bars do not exceed 12 mm in diameter. In this case a cover of 25 mm can be used.

5.2 Axial loads

The permissible axial load N on a short column reinforced with longitudinal bars and lateral ties (in accordance with the provisions of NCP 1) should not exceed that given by:

$$N = p_{cc} A_c + p_{sc} A_{sc} \tag{5.1}$$

in which p_{cc} is the permissible stress for concrete in direct compression, A_c is the cross-sectional area of concrete (excluding any finishing material and reinforcing steel), p_{sc} is the permissible compressive stress for the column bars, and A_{sc} is the cross-sectional area of the longitudinal steel.

To obtain the permissible axial load on a long reinforced concrete column the appropriate permissible load for a short column [as given by equation (5.1)] is multiplied by a factor termed the reduction coefficient. The reduction coefficient (a result of experimental and analytical work on long columns) is a function of the slenderness ratio (the ratio of the effective length of the column to its least lateral dimension). The reduction coefficients α for varying values of the slenderness ratios β can be obtained from the expressions:

$$\alpha = (45 - \beta)/30\,;\ 15 \leq \beta \leq 33 \tag{5.2}$$

$$\alpha = (57 - \beta)/60\,;\ 33 \leq \beta \leq 57 \tag{5.3}$$

Example 5.1

An internal column (225 mm × 225 mm) supports a symmetrical arrangement of beams. The total reaction from the beams, including the self-weight of the column, has been estimated to be 550 kN. Determine the amount of longitudinal reinforcement required, and the pitch and diameter of the transverse reinforcement. Assume $p_{cc} = 5.3\ \text{N/mm}^2$ and $p_{sc} = 112\ \text{N/mm}^2$.

Solution

A_c = the cross-sectional area of concrete $= 225^2 = 50\,625\ \text{mm}^2$

Putting $N = 550$ kN in equation (5.1) gives the result:

$$A_{sc}/A_c = (550 \times 10^3 - 5.3 \times 50\,625)/50\,625 \times 112$$
$$= 0.0497$$
$$A_{sc} = 2515\ \text{mm}^2$$

From Table 3.1: provide eight 20 mm ϕ bars (2513 mm²)

The pitch of the transverse reinforcement is the least of:

- 225 mm
- 12 × 20 = 240 mm
- 300 mm

Therefore 225 mm is the required pitch.

The Code provides that the diameter of the transverse reinforcement should not be less than one-quarter of the diameter of the main reinforcement and in no case less than 5 mm.

Therefore, use 6 mm ϕ bars at 225 mm pitch for the transverse reinforcement.

5.3 Bending moments

The decision of whether or not a column should be designed for bending moments depends on the location within the entire structure, and the nature of the adjoining beams, slabs and columns. For most internal columns supporting a symmetrical arrangement of beams and loading, the bending moments need not be taken into account. However, for external columns and for internal ones that are asymmetrically loaded, special consideration must be given to the bending moments.

The moment imposed on a particular column depends on the number of bays of which the column is a part and the stiffness coefficients of the adjoining beams and columns. A bay is essentially a system of frames formed by columns and beams. In Fig. 5.1, columns A and C belong to a one-bay system, whereas column B is part of a two-bay system.

The stiffness factor K_f is the ratio of the product of the elastic modulus E and the second moment of area I, to the effective length L of a structural element. Thus:

$$K_f = EI/L \tag{5.4}$$

Table 11 of NCP 1 provides formulae for calculating the bending moments in external columns, and in internal columns supporting an asymmetrical arrangement of beams and loading. The formulae are presented here in Table 5.2, in which:

M_e = bending moment at the end of the beam framing into the column, assuming fixity at both ends of the beam;

Fig. 5.1 Definition of bays

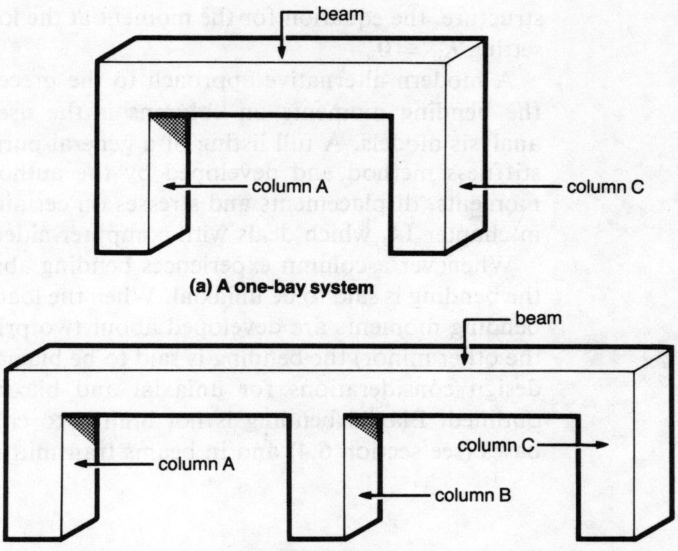

(a) A one-bay system

(b) A two-bay system

M_{es} = the maximum difference between the moments at the ends of the two beams framing into opposite sides of the column, each calculated on the assumption that the ends of the two beams frame into opposite sides of the column and each calculated on the assumption that the ends of the beams are fixed and one of the beams is unloaded;

K_b = the stiffness of the beam;

K_{b1} = the stiffness of the beam on one side of the column;

K_{b2} = the stiffness of the beam on the other side of the column;

K_l = the stiffness of the lower column; and

K_u = the stiffness of the upper column.

For prismatic members the stiffness factor (for the purpose of Table 5.2), can be obtained by simply dividing the moment of inertia of the cross-section by the length. In the case of columns in the top-most storey of a

Table 5.2 Bending moments in columns

		Moments for frames of one bay	Moments for frames of two or more bays
External and similarly loaded columns	Moments at foot of upper column	$\left(\frac{K_u}{K_l + K_u + 0.5K_b}\right)M_e$	$\left(\frac{K_u}{K_l + K_u + K_b}\right)M_e$
	Moments at head of lower column	$\left(\frac{K_l}{K_l + K_u + 0.5K_b}\right)M_e$	$\left(\frac{K_l}{K_l + K_u + K_b}\right)M_e$
Internal columns	Moment at foot of upper column	—	$\left(\frac{K_u}{K_l + K_u + K_{b1} + K_{b2}}\right)M_{es}$
	Moments at head of lower column	—	$\left(\frac{K_l}{K_l + K_u + K_{b1} + K_{b2}}\right)M_{es}$

structure, the equation for the moment at the lower column may be used by setting $K_u = 0$.

A modern alternative approach to the preceding method of calculating the bending moments on columns is the use of computer-based frame analysis models. A full listing of a general-purpose program, based on the stiffness method and developed by the author for analysing the forces, moments, displacements and stresses on certain plane frames, is presented in chapter 14, which deals with computer-aided design.

Whenever a column experiences bending about only one principal axis, the bending is said to be uniaxial. When the loading conditions are such that bending moments are developed about two principal axes (one major and the other minor) the bending is said to be biaxial. In section 5.4, the special design considerations for uniaxial and biaxial bending in columns are outlined. Biaxial bending is not limited to columns; it can also occur in bases (see section 6.4) and in beams transmitting eccentric thrusts.

5.4 Combined thrusts and moments

Thrusts are compressive direct forces. In studying the effects of combined thrusts and moments on columns, two different cases must be identified:

(1) the entire column section is in compression;

(2) a portion is in tension, while the rest is in compression.

In the second case the tensile resistance in the concrete is generally ignored if the maximum tensile stress developed is more than 10 per cent of the permissible compressive stress. It is always expedient to replace any applied moment by an equivalent system comprising the applied thrust acting at an eccentricity e from the centroid of the stressed area (which is not necessarily the centroid of the unreinforced section).

The following discussion is limited to short columns. Uniaxial and biaxial bending are treated separately.

5.4.1 Uniaxial bending

From elementary principles of the mechanics of materials it is known that, at a given section in an elastic member subjected to bending, the following expression must be valid:

$$M/I = \sigma_b/y = E/R \qquad (5.5)$$

where M is the bending moment, I is the second moment of area of the section, σ_b is the stress (due to bending only) at a distance y from the centroid, E is the modulus of elasticity (of the material), and R is the radius of curvature of the section. Thus:

$$\sigma_b = My/I \qquad (5.6)$$

in which σ_b may be positive (compressive) or negative (tensile), depending on y.

The applied moment M is equivalent to the axial thrust N acting at an eccentricity e such that:

$$M = Ne$$

or

$$e = M/N \qquad (5.7)$$

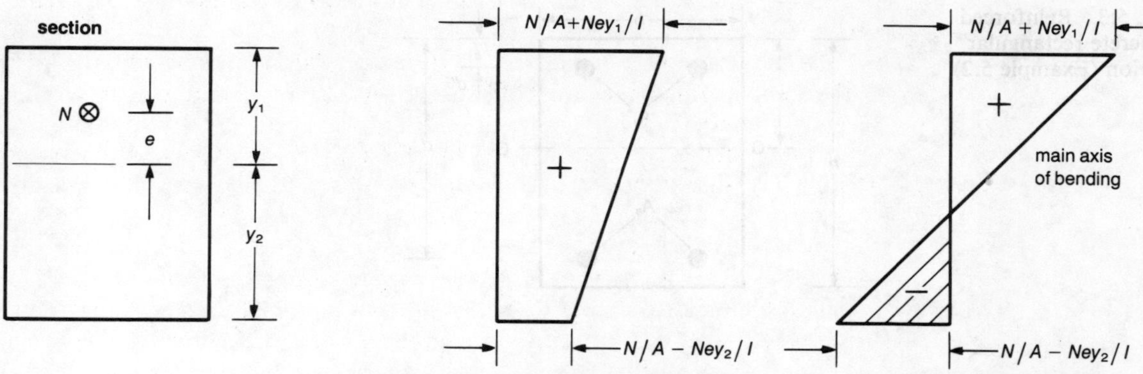

(a) Entire section in compression (b) Part of section in tension

Fig. 5.2 Stress distribution in column subjected to combined thrust and moment

If A is the area of the cross-section (A = equivalent concrete area if the section is reinforced), then the compressive stress due to the thrust is:

$$\sigma_a = N/A \tag{5.8}$$

Therefore the total stress σ (due to both axial thrust and moment) is given by:

$$\sigma = \sigma_a \pm \sigma_b \tag{5.9}$$

or, with equations (5.6) (5.7) and (5.8):

$$\sigma = N/A \pm Ne/I \tag{5.10}$$

where I must be taken as the equivalent second moment of area for a reinforced concrete section. The term 'equivalent' here means that the steel has been 'transformed' to an area of concrete that is capable of carrying the same amount of force as the steel being transformed (see Example 5.2).

Plots of σ versus y are shown in Fig. 5.2 for a rectangular section.

The maximum compressive and/or tensile stresses for any section must occur on the edges ($y = \pm y_{max}$). Thus:

$$\sigma_{max} = N/A \pm Ne/(I/y_{max}) \tag{5.11}$$

or, with $I/y_{max} = J$ (the section modulus):

$$\sigma_{max} = N/A \pm Ne/J \tag{5.12}$$

Therefore the design process must ensure that the maximum compressive stress in concrete σ_c does not exceed the permissible amount p_c. That is:

$$\sigma_c = N/A + Ne/J \leq p_c \tag{5.13}$$

Having derived equations (5.10)–(5.12), the analysis of any given section follows the pattern used for members subjected to bending only. The major difficulty in reinforced concrete members subjected to combined thrusts and moments (whether beams or columns) is the determination of A and J, which involves transforming the reinforcement to equivalent concrete and then finding the area and second moment of area of the resulting 'transformed' section.

The following example illustrates the application of the foregoing to a simple reinforced concrete section (see also Exercises 5.3–5.5).

Fig. 5.3 Reinforced concrete rectangular section (Example 5.2)

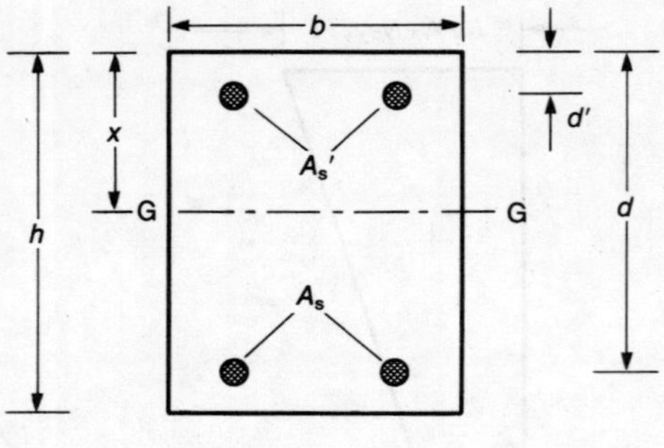

Example 5.2

Obtain an expression for the stress distribution in the rectangular reinforced concrete section (Fig. 5.3) which is to transmit an axial load N and a bending moment M about the axis G-G. Under what conditions will the entire section be just in compression?

Solution

The first part of the problem is to find an equivalent concrete section. Now if ϵ is the strain in the reinforcement with area A_s, then the force carried by the reinforcement is $\epsilon E_{st} A_s$, where E_{st} = the elastic modulus of the reinforcement.

The concrete that will replace the steel at the same location must obviously suffer the same strain ϵ. If the area of this concrete is A_{eq}, then the force carried by it is $\epsilon E_c A_{eq}$, where E_c is the elastic modulus of the concrete. Thus, for the two to be equivalent:

$$\epsilon E_{st} A_s = \epsilon E_c A_{eq}$$

or $$A_{eq} = mA_s \tag{5.14}$$

where $m = E_{st}/E_c$ = the modular ratio. Similarly the equivalent concrete area for the reinforcement with area A_s' is:

$$A_{eq}' = mA_s' \tag{5.15}$$

The transformed section will be as shown in Fig. 5.4. Therefore the area A of the transformed section is:

$$A = bh + m(A_s + A_s') - (A_s + A_s') \tag{5.16}$$

the last term being necessary in order to account for the concrete area already displaced by the reinforcement. Thus:

$$A = bh + (m-1)(A_s + A_s') \tag{5.17}$$

By taking first moments of area about the top of the section:

$$A\bar{x} = bh(h/2) + mA_s(d) + mA_s(d') - A_s(d) - A_s'(d')$$

or

$$\bar{x} = [bh^2/2 + (m-1)(A_s d + A_s' d')]/A \tag{5.18}$$

By taking second moments of area about axis G-G:

$$I_{GG} = \int_{-(h-\bar{x})}^{\bar{x}} by^2 dy + mA_s(d-\bar{x})^2 + mA_s'(\bar{x}-d')^2$$
$$- A_s(d-x)^2 - A_s'(\bar{x}-d')^2$$

Fig. 5.4 Transformed section (solution to Example 5.2)

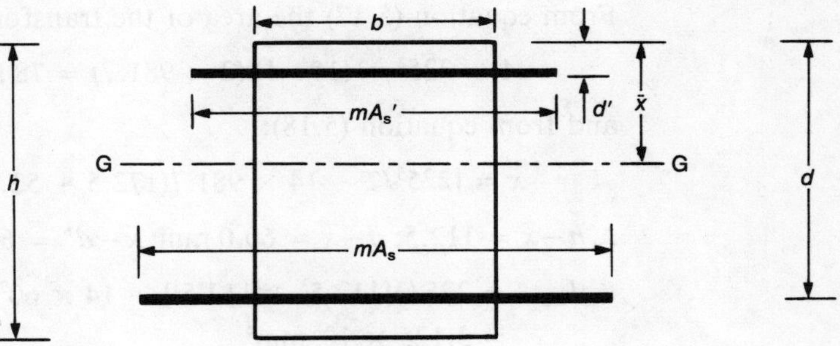

or

$$I_{GG} = b/3[\bar{x}^3 + (h-\bar{x})^3] + (m-1)[A_s(d-\bar{x})^2 + A_s'(\bar{x}-d')^2] \quad (5.19)$$

Using equation (5.10) with e = M/N the stress distribution is given by:

$$\sigma = N/A + Ney/I_{GG} \qquad (d-\bar{x}) \leq y \leq \bar{x} \quad (5.20)$$

where A and I_{GG} are given by equations (5.17) and (5.19), respectively. Equation (5.20) constitutes the required expression.

If the section is to be just in compression, it is required that $y = -(d-\bar{x})$, so:

$$\sigma_{min} = N/A - Ne(d-\bar{x})/I_{GG} = 0 \quad (5.21)$$

That is:

$$e = M/N = I_{GG}/A(d-\bar{x})$$

or

$$M/N = \frac{b/3[\bar{x}^3 + (h-\bar{x})^3] + (m-1)[A_s(d-\bar{x})^2 + A_s'(\bar{x}-d')^2]}{[bh + (m-1)(A_s + A_s')](d-\bar{x})}$$

Note that if the section is not reinforced, $A_s = A_s' = 0$ and $A = bh$; $\bar{x} = h/2$; and $I_{GG} = bh^3/12$. In this case if the entire section is to be always in compression:

$$\sigma_{min} = N/bh - Ne(\tfrac{1}{2}h)/(bh^2/12) \geq 0 \quad (5.23)$$

or

$$e = M/N \leq h/6 \quad (5.24)$$

This constitutes an important result in the design of members subjected to eccentric loading, or combined thrusts and moments.

Example 5.3

A 225 mm × 225 mm reinforced concrete column is required to transmit an axial load of 75.8 kN. What is the maximum bending moment the column can carry (in addition to the axial load) without developing tensile stresses, if it is to be reinforced with four 25 mm ϕ bars? In the limiting condition, what is the maximum stress developed in the concrete? Assume $m = 15$.

Solution

$b = h = 225$ mm; $A_s = A_s' = 2 \times \pi(25^2)/4 = 981.7$ mm²; $N = 75.8$ kN; $d' = 40 + 12.5 = 52.5$ mm; and $d = h - 52.5 = 172.5$ mm.

From equation (5.17) the area of the transformed section is:

$$A = 225^2 + (15-1)(2 \times 981.7) = 78\,114\,\text{mm}^2$$

and from equation (5.18):

$$\bar{x} = [225^3/2 + 14 \times 981.7(172.5 + 52.5)]/78\,114 = 112.5\text{ mm}$$

$$\therefore h-\bar{x} = 112.5;\ d-\bar{x} = 60.0\text{ mm};\ \bar{x}-d' = 60.0\text{ mm}$$

$$\therefore I_{GG} = 225/3[112.5^3 + 112.5^3] + 14 \times 60^2 \times 2 \times 981.7$$
$$= 3.125 \times 10^8\text{ mm}^4$$

For the limiting condition, $\sigma_{min} = 0$, and from equation (5.21):

$$e = \frac{M}{N} = I_{GG}/A(d-\bar{x}) = 3.125 \times 10^8/(78\,114 \times 60) = 66.68\text{ mm}$$

or

$$M = 66.68 \times 75.8/10^3 = 5.1\text{ kN m}$$

The maximum stress in concrete is given by equation (5.20) with $y = \bar{x} = 112.5$ mm, ie:

$$\sigma_{max} = 75.8 \times 10^3(1/78\,114 + 66.68 \times 112.5/3.125 \times 10^8)$$
$$= \underline{2.8\text{ N/mm}^2}$$

5.4.2 Biaxial bending

In biaxial bending there are two principal axes about which bending takes place. The principal axes are the axes about which the second moment of area about the centroid achieves its maximum (major axis) and minimum (minor axis) values. The principal axes are orthogonal to each other and intersect at the centroid of the section.

Biaxial bending may arise as a result of moments applied directly on the

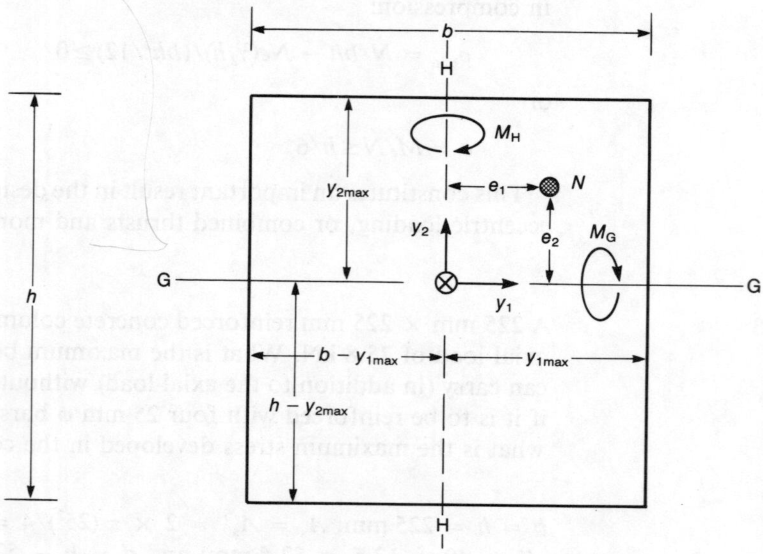

Fig. 5.5 Rectangular section carrying direct thrust N at eccentricit e from the centroid

member or may simply be due to an eccentric action of a direct thrust. From the point of view of analysis, the latter case is the more general because all classes of problems can be reduced to that of an equivalent system involving eccentric loading of the member.

Consider the rectangular section shown in Fig. 5.5 which carries a direct thrust N at an eccentricity $e = \sqrt{(e_1^2 + e_2^2)}$ from the centroid of the section.

The principal axes are G-G and H-H, and if the bending moments about these axes are M_G and M_H, respectively then:

$$M_G = Ne_1 \tag{5.25}$$

$$M_H = Ne_2 \tag{5.26}$$

The stresses due to these bending moments can be obtained, using equation (5.5), as:

$$\sigma_{bG} = M_G y_2/I_{GG} = Ne_2 y_2/I_{GG} \tag{5.27}$$

$$\sigma_{bH} = M_H y_1/I_{HH} = Ne_2 y_1/I_{HH} \tag{5.28}$$

As the compressive stress due to the direct thrust is N/A, the total stress at any point with coordinates y_1, y_2 (measured relative to the principal axes) is:

$$\sigma = N/A + Ne_2 y_2/I_{GG} + Ne_1 y_1/I_{HH} \tag{5.29}$$

in which A is the area of the cross-section. For a reinforced section, the area A and second moments of area I_{GG} and I_{HH} must be referred to the transformed section. The extreme stresses are developed at the corners of the section and the design process must ensure that these do not exceed the allowable amounts. Thus, the maximum compressive stress achieved when $y_1 = y_{1max}$ and $y_2 = y_{2max}$ is:

$$\sigma_{max} = N(1/A + e_2/J_{GG} + e_1 J_{HH}) \leq p_c \tag{5.30}$$

where $J_{GG} = I_{GG}/y_{2max}$ and $J_{HH} = I_{HH}/y_{1max}$ are the section moduli in the principal axes G-G and H-H, respectively.

On the other hand, it may be desirable to have the entire section always in compression, in which case:

$$\sigma_{min} = N(1/A - e_2/J_{GG}' - e_1/J_{HH}') \geq 0 \tag{5.31}$$

where $J_{GG}' = I_{GG}/(h - y_{2max})$ and $J_{HH}' = I_{HH}/(b - y_{1max})$

Equations (5.30) and (5.31) may then serve as conditions for selecting the section geometry (and therefore A, J_{GG} and J_{HH}).

5.5 Exercises

5.1 An internal column (400 mm × 250 mm × 2.5 m) supports a symmetrical arrangement of beams. The column is required to transmit a total thrust (not including its weight) of 672 kN. Determine the amount of longitudinal reinforcement required, plus the pitch and diameter of the transverse reinforcement. Assume p_{cc} = 5.3 N/mm^2 and p_{sc} = 112 N/mm^2.

5.2 If the architect now wants to replace the rectangular column of Exercise 5.1 with a circular one, determine the minimum diameter and the amount of longitudinal reinforcement required under the same loading conditions.

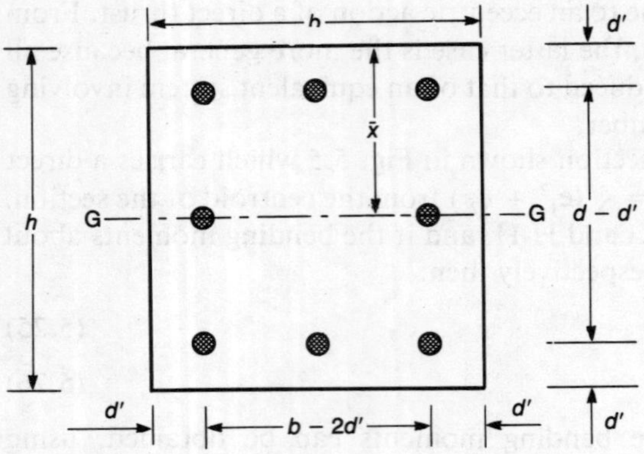

Fig. Q5.3(a)

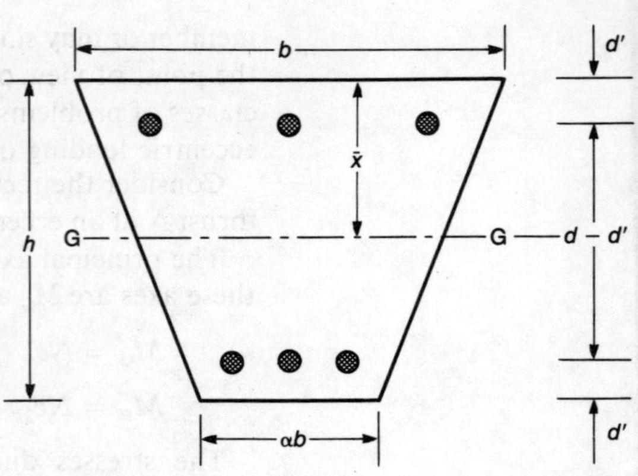

Fig. Q5.3(b)

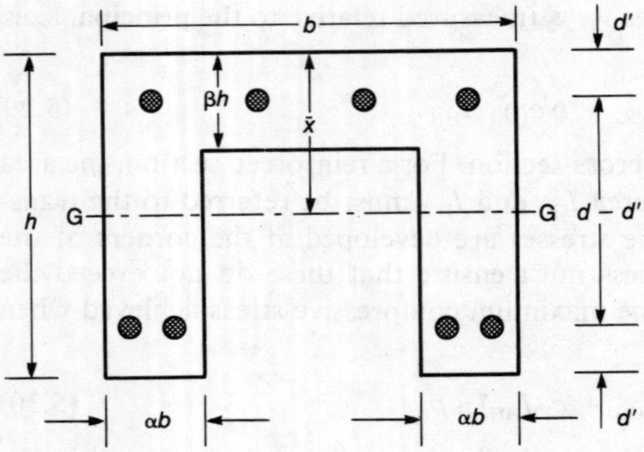

Fig. Q5.3(c)

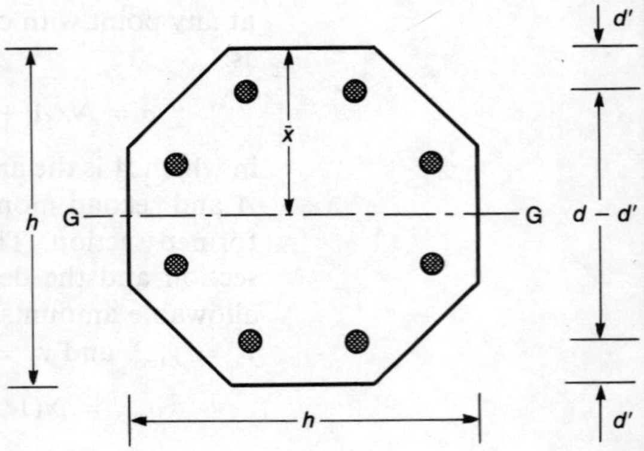

Fig. Q5.3(d)

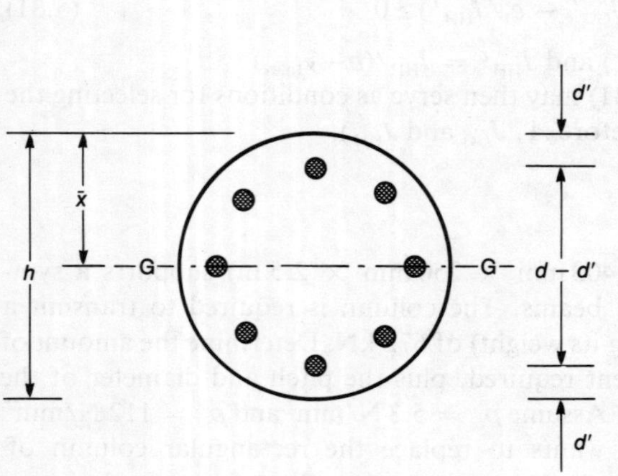

Fig. Q5.3(e)

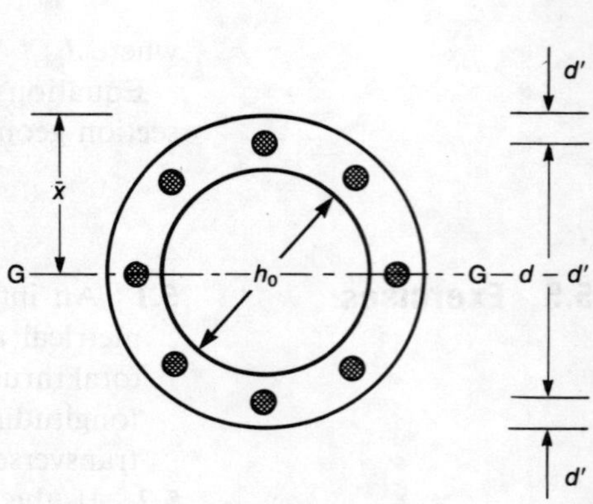

Fig. Q5.3(f)

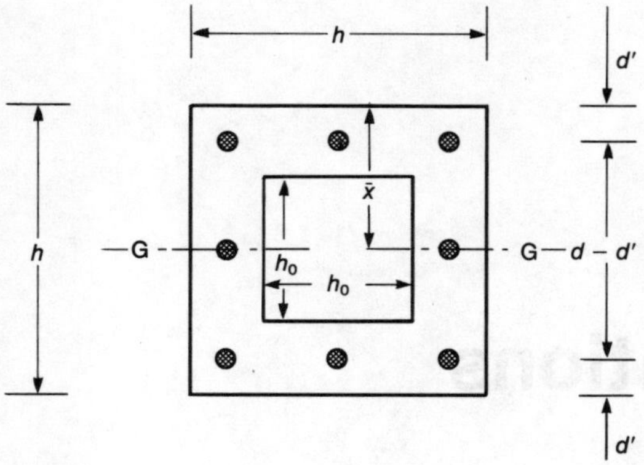

Fig. Q5.3(g)

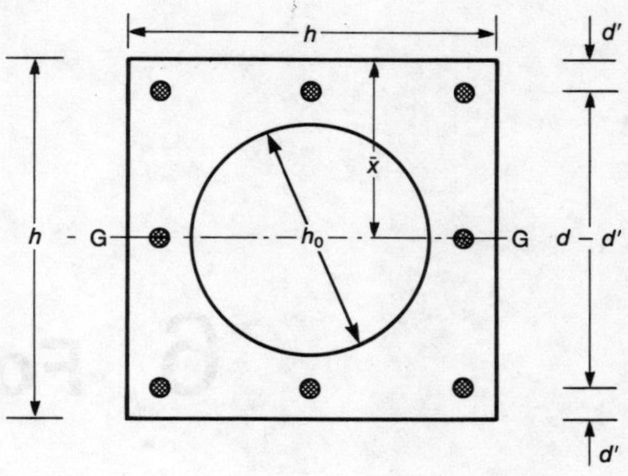

Fig. Q5.3(h)

5.3 Obtain expressions for the stress distribution in the reinforced concrete sections shown in Figs. Q5.3(a) to (h), given that bending takes place about axis GG only and the section carries a direct thrust N. All reinforcing bars have the same diameter.

5.4 From the results obtained in Exercise 5.3, find the conditions under which the various sections will be just in compression.

5.5 Determine the moment of resistance of each of the sections shown in Figs Q5.3(a) to (h), assuming that the concrete is not effective in tension.

5.6 A hollow square column (exterior dimensions 400 mm × 400 mm, interior dimensions 150 mm × 150 mm) is reinforced with four 20 mm diameter bars. The column is to transmit an axial load of 120.0 kN. What is the maximum bending moment that the column is capable of carrying? Assume m = 15.

5.7 What is the maximum stress developed in the column of Exercise 5.6 if the section is to be just in compression?

5.8 Attempt Exercise 5.6 again, assuming the central hollow square is replaced with a circular duct of diameter 150 mm. What will the maximum compressive stress under the limiting condition be now?

6 Foundations

6.1 Introduction

As defined in section 1.2.4, the foundation of any structure is a base that allows the load from the structure to be transmitted to the soil in the most efficient way. The design process for any given base must ensure that the load distribution on the soil is such that the load-bearing capacity of the soil is not exceeded.

Ideally, before a foundation is designed for any structure, a detailed soil investigation of the building site should be performed. The nature of the investigation required is dependent on the site conditions and the type of structure to be erected. For most common structures (eg, buildings less than three storeys high) on relatively firm ground, an experienced engineer may often be able to perform a visual investigation and arrive at a safe estimate of the load-bearing capacity. However, actual soil investigation is advisable before the design of any foundation.

Table 6.1 Permissible soil pressures for some common soil types

Description of soil	Bearing capacity (kN/m^2)
Silt	75
Soft clay	150
Stiff clay	250
Fine/loose sand	200
Coarse/loose sand	300
Loose gravel	400
Compact sand-gravel mixture	600
Schist/slate rocks	4000
Granite/bedrock	10 000

In Table 6.1 permissible soil pressures are given for some common soil types and these can be used when better on-site estimates are not available.

The design considerations for isolated bases, combined bases and rafts are outlined in the following sections.

6.2 Isolated bases

An isolated base transmits both direct thrusts and bending moments. The moments may arise from the non-axial (or eccentric) nature of the direct thrusts, or may simply be due to the bending moments developed at the base of the columns.

In general, therefore, it is expedient to consider the isolated base as transmitting a direct thrust N acting at an eccentricity e from the centroid of the base, such that $e = M/N$ (M being the moment being transmitted). As shown for columns under the combined influence of direct thrusts and moments, the stress distribution at the base may be either trapezoidal or triangular, depending on the relative values of the eccentricity e and the dimensions of the base.

6.2.1 Concentrically-loaded isolated bases

The simplest case occurs when there is no bending moment to be transmitted and/or the load is concentric ($e = 0$). The stress distribution will then be rectangular. For a rectangular base, values of e less than one-sixth [see equation (5.24)] of the lateral dimension of the base will produce a trapezoidal stress distribution. When e exceeds the critical value, the stress distribution is triangular and, for a base, this simply implies that some portions of the soil are unloaded. The latter situation must be avoided as much as possible.

If N = concentric load (including base self-weight) on the base and q = permissible soil-bearing pressure, then the minimum base area required is N/q. Now, consider a rectangular base (Fig. 6.1) of length L and width B. The net ground pressure acting on the base is $f = N/BL$ and B and L must be chosen such that $f \leq q$. That is:

$$N/BL \leq q \tag{6.1}$$

or

$$BL \geq N/q \tag{6.2}$$

If the rectangular base supports a rectangular column of dimension $l \times b$ the critical sections for bending moments and shear are as shown in Fig. 6.1 (*cf*. Fig. 4.5 and Fig. 4.6 for flat slabs).

The maximum bending moment about the H-axis can be obtained by considering the moment due to the load in the shaded portion on Fig. 6.1(a) overleaf. As the area of the shaded portion is $B(L-l)/2$ the total load acting on the area is $fB(L-l)/2$. Thus, the maximum bending moment about the H-axis is:

$$\begin{aligned} M_H &= fB(L-l)/2 \times {}^1\!/_2[(L-l)/2] \\ &= fB(L-l)^2/8 \end{aligned} \tag{6.3}$$

Similarly:

$$M_G = fL(B-b)^2/8 \tag{6.4}$$

For a base supporting a circular column all that is required is to set $l = b = D$, where D is the diameter of the column.

To obtain the required base thickness for bending, the procedure followed for a beam under flexure can be used. It is not always necessary to provide a uniform base thickness, as the moment and shear are maximum in the neighbourhood of the column and decrease to zero at the base edges. The base can be tapered as shown in Fig. 6.1(b).

It is always important to ensure that the thickness provided (using

Fig. 6.1 Rectangular isolated base

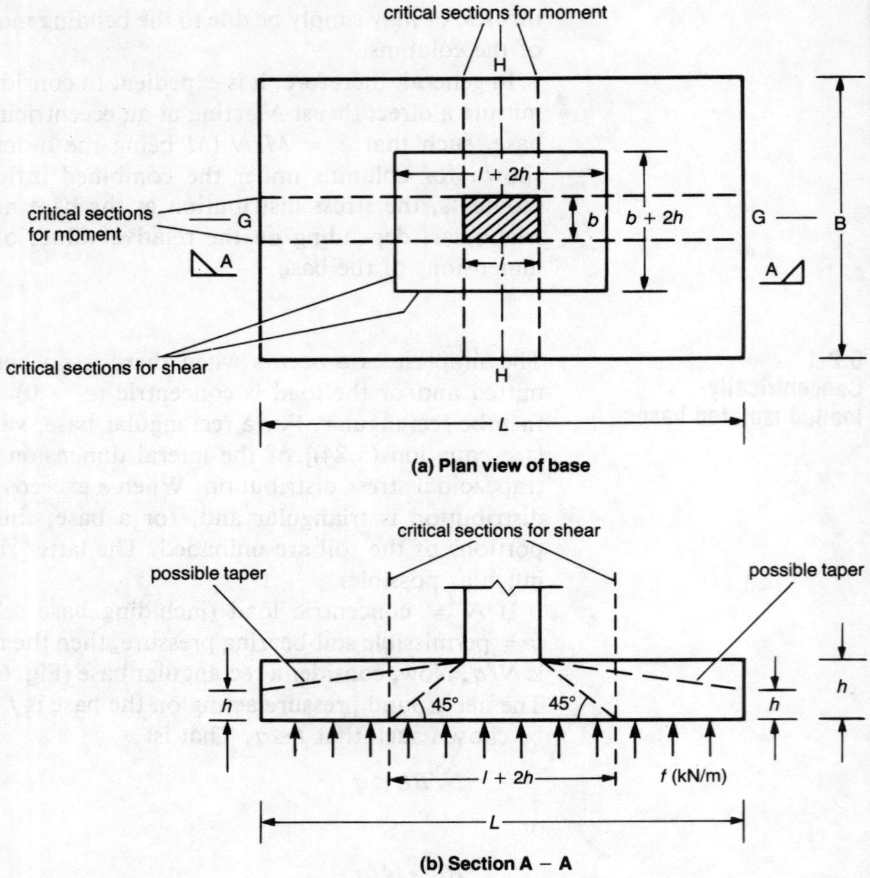

bending considerations) is adequate to meet the requirements of punching shear (see section 1.3.2). The permissible punching shear stress in concrete is about 20 per cent of the permissible compressive stress in bending. Usually, the minimum thickness for punching is used as a first estimate of the base weight.

Example 6.1

A square reinforced concrete base is to support a central concrete column (350 mm square) which carries a total load of 500 kN. If the allowable soil pressure is 120 kN/m², determine the appropriate base size and the required reinforcement. Assume $p_{st} = 126\ \text{N/mm}^2$, $p_c = 7\ \text{N/mm}^2$ and $m = 15$.

Solution

Concentric load $N = 500$ kN; allowable soil pressure $q = 120\ \text{kN/m}^2$; and permissible punching shear stress $= 0.2 \times 7 = 1.4\ \text{N/mm}^2$. Therefore:

minimum base thickness for punching

$$= \frac{\text{load}}{\text{perimeter of punching surface} \times \text{punching shear stress}}$$

$$= 500 \times 10^3/(4 \times 350 \times 1.4)$$

$$= 255\ \text{mm, say } 260\ \text{mm}$$

Assuming the density of concrete = 2440 kg/m^3, then the pressure on the soil due to the weight of the base = $0.26 \times 2440 \times 9.81 \times 10^{-3}$ = 6.22 kN/m^2.

Therefore, the net soil pressure = 120 − 6.22 = 113.78 kN/m^2 and the minimum base area required = 500/113.78 = 4.4 m^2.

[An alternative to the above computations is to assume the area of base required = Am^2. Then the base self-weight = $6.22A$ kN. Thus, the load on the soil = $500 + 6.22A$ and the minimum base area must be such that $A = (500 + 6.22A)/120$, or $A = 4.4$ m^2.]

If B is the base side, then the base area is:

$$B^2 \geq 4.4\,\text{m}^2 \text{ or } B \geq 2.1\,\text{m}$$

Therefore, provide a 2.2 m × 2.2 m base.

The soil pressure on the base = $f = (500 + 6.22 \times 2.2^2)/2.2^2$

$$= 109.5\,\text{kN/m}^2$$

The maximum bending moment on the base, using equation (6.3), is $M = 109.5 \times 2.2 \times (2.2 - 0.35)^2/8 = 103.1$ kN m.

$$\text{Effective depth required} = \sqrt{(103.1 \times 10^3/1.34 \times 2.2)}$$
$$= 187\,\text{mm}.$$

Assuming 60 mm total cover, the minimum base thickness required is $h = 187 + 60 = 247$ mm.

Hence the 260 mm thickness obtained when punching is taken into consideration is still acceptable.

Next, it is necessary to ensure that the thickness provided is adequate for resistance against direct shear. The periphery of the critical section for shear (see Fig. 6.1) is $b_0 = 4 \times (350 + 2 \times 260) = 3480$ mm and the area enclosed is $A_0 = (0.35 + 2 \times 0.26)^2 = 0.76$ m^2. Thus the direct shear at the critical section is:

$$V = 109.5 \times [2.2^2 - 0.76] = 447.1\,\text{kN}$$

Using equation (4.5), the nominal shear stress is:

$$V = 447.1 \times 10^3/(0.85 \times 200 \times 3480) = 0.75\,\text{N/mm}^2$$

which is less than the permissible amount of $\sqrt{7} = 2.6$ N/mm^2. Therefore a base thickness of 260 mm is acceptable.

$$\text{Effective depth} = 260 - 60 = 200\,\text{mm}$$

The value of the required A_{st} is given by:

$$A_{st} = M/p_{st}\,z = 103.1 \times 10^6/126 \times 0.85 \times 200)$$
$$= 4813\,\text{mm}^2$$

Therefore, use sixteen 20 mm ϕ bars each way (5027 mm^2).

Bond: At this juncture, because of its importance in the design of bases, the special consideration required for bond will be outlined. One of the important assumptions in the modular-ratio method is that there exists a perfect bond between concrete and steel. The assumption of a perfect bond is not limited to the elastic method as will be seen in later chapters. Bond is, essentially, resistance to slippage. The magnitude of this resistance per unit area of action is what is termed a bond stress.

The requirements of NCP 1 are that the most critical local bond stress be calculated at the section of maximum bending moment, ie, (for this problem) at the face of the column. The local bond stress v_b, is given by (see NCP 1, clause 3.11.2)

$$v_b = V/zO \tag{6.5}$$

in which V is the total shear across the section, z is the arm of the resistance moment, and O is the sum of the perimeters of the tensile reinforcement bars.

At the face of the column:

$$V = 109.5 \times (2.2 - 0.35) \times 2.2/2 = 222.8 \text{ kN}$$

and

$$v_b = 222.8 \times 10^3/(0.85 \times 200 \times \pi \times 20 \times 16)$$
$$= 1.3 \text{ N/mm}^2$$

which is greater than the allowed maximum value of 1.25 N/mm² (see Table 2.1). An attempt must now be made to increase the sum of the perimeters by choosing smaller bar diameters and an increased number of reinforcing bars. Therefore, try twenty-five 16 mm ϕ bars (4825 mm²) each way. In which case, by simple proportion with the calculated value above:

$$v_b = 20 \times 16 \times 1.3/25 \times 16 = 1.04 \text{ N/mm}^2$$

which is acceptable.

Therefore, provide a 2.2 m × 2.2 m × 260 mm base with twenty-five 16 mm ϕ bars (bottom) each way.

6.2.2 Eccentrically-loaded isolated bases

Consider a rectangular base which carries a direct thrust N, at an eccentricity e. As shown earlier for eccentrically-loaded columns (section 5.4), the two possible pressure distributions are as given in Fig. 6.2. The difference is that, unlike the case of the columns, no tensile stresses are developed here, the absence of compressive stresses being indicative of the unloaded

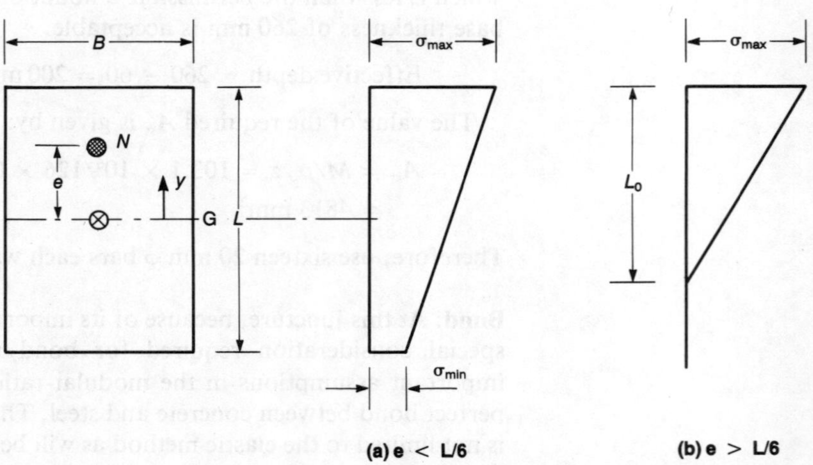

Fig. 6.2 Stress distributions on eccentrically-loaded bases

portion of the soil. Therefore, the situation depicted by Fig. 6.2(b) should be avoided as much as possible, because the soil load-bearing capacity is not being fully utilised.

From the previous analysis on eccentrically-loaded columns and for the case in Fig. 6.2(a) ($e<L/6$), the maximum compressive stress is:

$$\sigma_{max} = N/BL + Ne/J_{GG} \tag{6.6}$$

in which J_{GG} [$= (1/12)\,BL^3/(L/2) = BL^2/6$] is the section modulus. That is:

$$\sigma_{max} = N/BL(1 + 6e/L) \tag{6.7}$$

Similarly:

$$\sigma_{min} = N/BL(1 - 6e/L) \tag{6.8}$$

showing that $e = L/6$ for the limiting condition when $\sigma_{min} = 0$.

For the case depicted by Fig. 6.2(b) (ie, $e>L/6$), the equilibrium of all vertical forces on the base requires that:

$$N = {}^1\!/_2\, \sigma_{max}\, BL_0 \tag{6.9}$$

By taking moments of all forces about the maximum compression face:

$$N(L/2 - e) = {}^1\!/_2\, \sigma_{max}\, BL_0\, ({}^1\!/_3 L_0) \tag{6.10}$$

or, with equation (6.9):

$$L_0 = 3(L - 2e)/2 \tag{6.10}$$

Eliminating L_0 in equation (6.9), the expression for the maximum compressive stress is:

$$\sigma_{max} = 4N/[3B(\mathrm{L} - 2e)] \tag{6.11}$$

The design process must ensure that the base area is such that:

$$\sigma_{max} \not> \mathrm{q} \tag{6.12}$$

where q is the allowable soil-bearing capacity.

Once the appropriate base area has been selected the new net soil-pressure distribution can be computed, and the base designed in the same way as for the earlier case of concentric loading. However, the loading for this particular case is no longer uniformly-distributed but is either trapezoidal ($e<L/6$) or triangular ($e \geq L/6$).

Example 6.2

A 3 m long rectangular base is required to support (somewhere along the middle of its width) a reinforced concrete column (400 mm square) which carries a combined thrust of 650 kN and a moment of 280 kN m (see Fig. 6.3).

Determine the appropriate base width, base thickness and reinforcement, given that the allowable soil pressure is 110 kN/m^2. Assume $p_{st} = 126\ \mathrm{N/mm^2}$, $p_c = 7\ \mathrm{N/mm^2}$ and $m = 15$.

Solution

The effect of the moment is that the base carries an eccentric load of 650 kN at an eccentricity $e = 280/650 = 0.43$ m.

As $L/6 = 3/6 = 0.5 > e$, the stress distribution will be akin to that

Fig. 6.3 Solution to Example 6.2

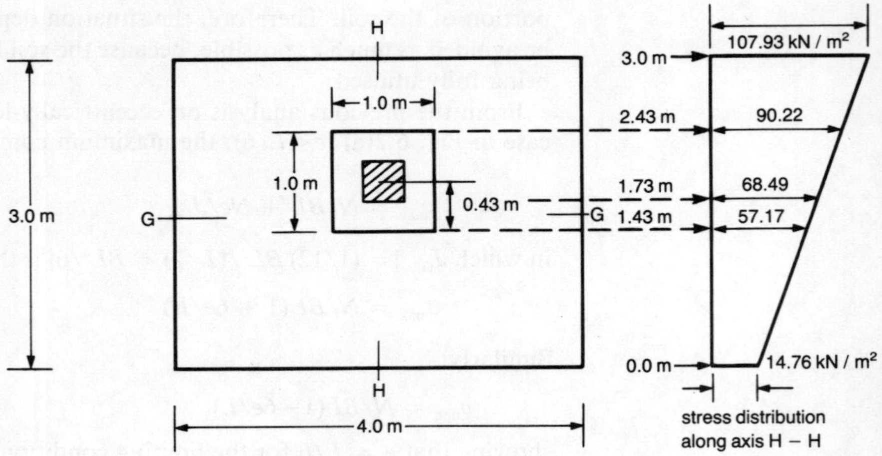

shown in Fig. 6.2(a). Using equation (6.7) the maximum stress due to the column load only (without base self-weight) is:

$$\sigma_{max} = 650(1 + 6 \times 0.43/3)/3B = 403/B \text{ kN/m}^2$$

The minimum base thickness required for punching is 650 × 10³/(4 × 400 × 1.4) = 290 mm, say 300 mm.

The intensity of load on the soil due to the weight of the base alone is 0.3 × 2440 × 9.81 × 10⁻³ = 7.18 kN/m². Therefore the net soil pressure = 110 − 7.18 = 102.82 kN/m² and:

$$\sigma_{max} = 403/B \leq 102.82 \text{ kN/m}^2$$

giving

$$B \geq 403/102.82 = 3.9 \text{ m}$$

Therefore, provide a 4 m wide base.

The maximum (total) soil stress on the base is now:

$$f_{max} = (403/4 + 7.18) = 107.93 \text{ kN/m}^2$$

and the minimum (total) stress is:

$$f_{min} = 650(1 - 6 \times 0.43/3)/4 \times 3 + 7.18 = 14.76 \text{ kN/m}^2$$

By taking moments about the faces of the columns:

$$M_G = 4[1.73 \times 14.76 \times 1.73/2 + 1.73 \times (68.49 - 14.76) \times 1.73/3] = 302.8 \text{ kN m}$$

$$M_H = \tfrac{1}{2}(14.76 + 107.93) \times 3 \times (2 - 0.2)^2/2 = 298.1 \text{ kN m}$$

The value of d required for M_G is given by:

$$M_G = \sqrt{\left\{\frac{302.8 \times 10^6}{1.34 \times 4 \times 10^3}\right\}} = 238 \text{ mm}$$

and the value of d required for M_H is given by:

$$M_H = \sqrt{\left\{\frac{298.1 \times 10^6}{1.34 \times 3 \times 10^3}\right\}} = 272.3 \text{ mm}$$

That is, M_H is the more critical. With 60 mm cover, the minimum thickness for bending is $h = 272.3 + 60 = 332.3$ mm.
Therefore, provide a base 340 mm thick.

(**Note:** This is slightly more than the 300 mm assumed thickness, based on punching considerations and used in computing the base self-weight. However, the effect on the above computation would be insignificant.)

It is now necessary to check for direct shear. The critical shear section is a 1 m square portion as shown in Fig. 6.3. The total shear at the periphery is:

$$V = \tfrac{1}{2}(107.93 + 14.76) \times 3 \times 4 - \tfrac{1}{2}(90.22 + 59.17) \times 1 \times 1$$
$$= 661.4 \text{ kN}$$

and the nominal shear stress is [see equation (4.5)]:

$$V = \frac{661.4 \times 10^3}{4 \times 10^3 \times 0.85(340-60)} = 0.7 \text{ N/mm}^2 < \sqrt{7} \text{ N/mm}^2$$

Therefore, the 340 mm base is acceptable for shear.

The value of A_{st} required for M_G is given by:

$$A_{st} = \frac{302.8 \times 10^6}{126 \times 0.85 \times 280} = 10\,097 \text{ mm}^2$$

Use thirty-three 20 mm ϕ bars bottom (10 369 mm^2) along the H-H axis.

The value of A_{st} required for M_H is given by:

$$A_{st} = 298.1 \times 10097/302.8 = 9940 \text{ mm}^2$$

Use thirty-two 20 mm ϕ bars bottom (10 054 mm^2) along the G-G axis.

Bond: The shear at the section of maximum moment is:

$$V = \tfrac{1}{2}(68.49 + 14.76) \times 4 \times 1.73 = 288 \text{ kN}$$

The bond stress, using equation (6.5), is:

$$v_b = \frac{288 \times 10^3}{(0.85 \times 280) \times 32 \times \pi \times 20} = 0.6 \text{ N/mm}^2 < 1.25 \text{ N/mm}^2$$

and is therefore acceptable.

6.3 Strip foundations

In designing a strip foundation, or any combined base for that matter, it is best to aim at a condition that will cause the centre of gravity of the loads and that of the base to coincide. If this ideal condition cannot be achieved then, at the very worst, the eccentricity introduced should be at most equal to one-sixth of the lateral dimension of the base. With reference to Fig. 6.4 overleaf the eccentricity is given by:

$$e = \frac{\Sigma N_j Y_j}{\Sigma N_j} \tag{6.13}$$

in which Σ is the summation for all column loads and $N_j Y_j$ is the moment of the jth thrust about the centre of gravity of the strip.

Once the eccentricity e has been determined, the pressure distribution can

Fig. 6.4 Loading on strip foundation

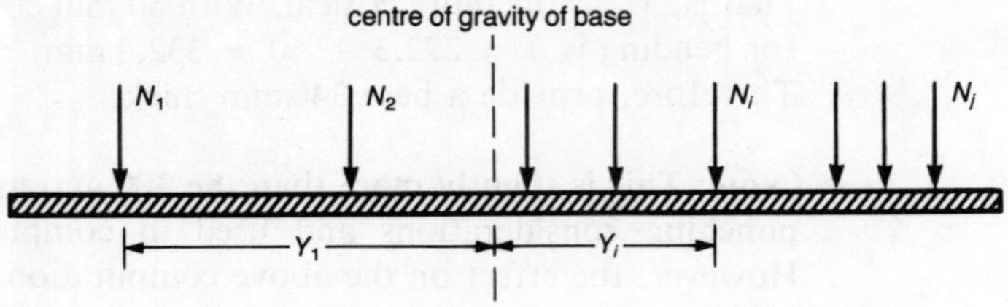

be obtained in the same way as for isolated bases. The base can then be designed as an up-turned continuous beam whose 'supports' are the columns (see Exercise 6.4).

6.4 Raft foundations

As in the design of other types of base, it is desirable, when designing a raft foundation, to have the centre of gravity of the loads coinciding with the centroid of the base. It is not always possible to achieve this condition.

For a rectangular raft, if the resultant load is eccentric, the extreme pressures are realized at the corners.

If e_1 and e_2 are the components of the eccentricity along the G-G and H-H axes respectively, as shown in Fig. 6.5, then equation (6.7) can be generalised for the four corners (1-4) to give:

$$\sigma_1 = [1 + 6e_1/B + 6e_2/L]\Sigma N_j/BL \tag{6.14}$$

$$\sigma_2 = [1 + 6e_1/B - 6e_2/L]\Sigma N_j/BL \tag{6.15}$$

$$\sigma_3 = [1 - 6e_1/B - 6e_2/L]\Sigma N_j/BL \tag{6.16}$$

$$\sigma_4 = [1 - 6e_1/B + 6e_2/L]\Sigma N_j/BL \tag{6.17}$$

in which $\sigma_i(i = 1,4)$ are the stresses (due to column loads only) at the corners; N_j is the jth load; Σ is the summation for all loads; and:

$$e_1 = B/2 - \Sigma N_j y_{ij}/\Sigma N_j \tag{6.18}$$

while

$$e_2 = L/2 - \Sigma N_j y_{ij}/\Sigma N_j \tag{6.19}$$

Fig. 6.5 Eccentrically-loaded rectangular raft foundation

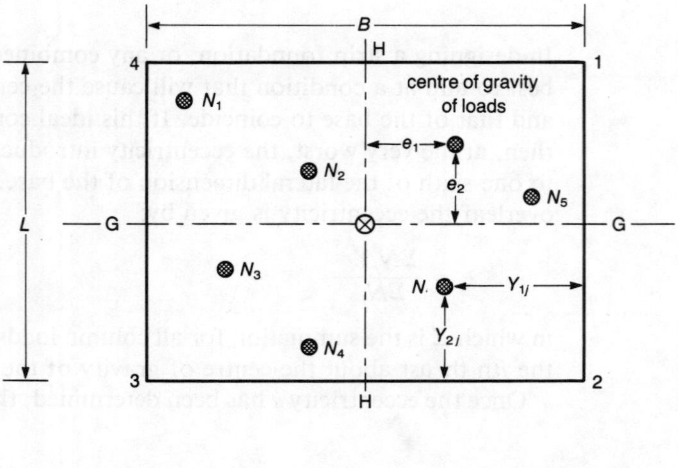

The maximum value of $\sigma_i (i = 1,4)$ given by equations (6.14) to (6.17) must not exceed the safe bearing pressure of the soil. The raft size (namely, B and L) is to be chosen on the basis of this criterion.

Once the net ground pressures are obtained the raft can be designed like an inverted floor slab. It must be emphasized that for effective use of the soil, the value of σ_i $(i = 1,4)$ must be at least equal to zero.

6.5 Exercises

6.1 A square concrete base is to support a concentric circular column (diameter = 400 mm) which transmits a total thrust of 895 kN. Given that the allowable soil pressure is 154 kN/m^2, determine the appropriate base size and the required reinforcement. Assume p_{st} = 140 N/mm^2, p_c = 7 N/mm^2 and m = 15.

6.2 If the width of the base in Exercise 6.1 is limited to 2.5 m, determine a suitable base length and reinforcement, assuming the same loading and site conditions.

6.3 Design a suitable base to support a concrete column (300 mm square) which carries a direct thrust of 720 kN and a moment of 342 kN m. The allowable soil pressure is 125 kN/m^2. Choose values for the permissible stresses, but assume m = 15.

6.4 Determine the suitable length, thickness and reinforcement for the strip base loaded as shown in Fig. Q6.4. The base length is not to exceed 35.5 m and the safe bearing pressure is 95 kN/m^2. The columns are 400 mm square. Assume p_{st} = 126 N/mm^2, p_c = 7 N/mm^2 and m = 15.

6.5 Redesign the base in Fig. Q6.4, given that the load carried by column number 8, on the extreme right, is 980 kN.

Fig. Q6.4

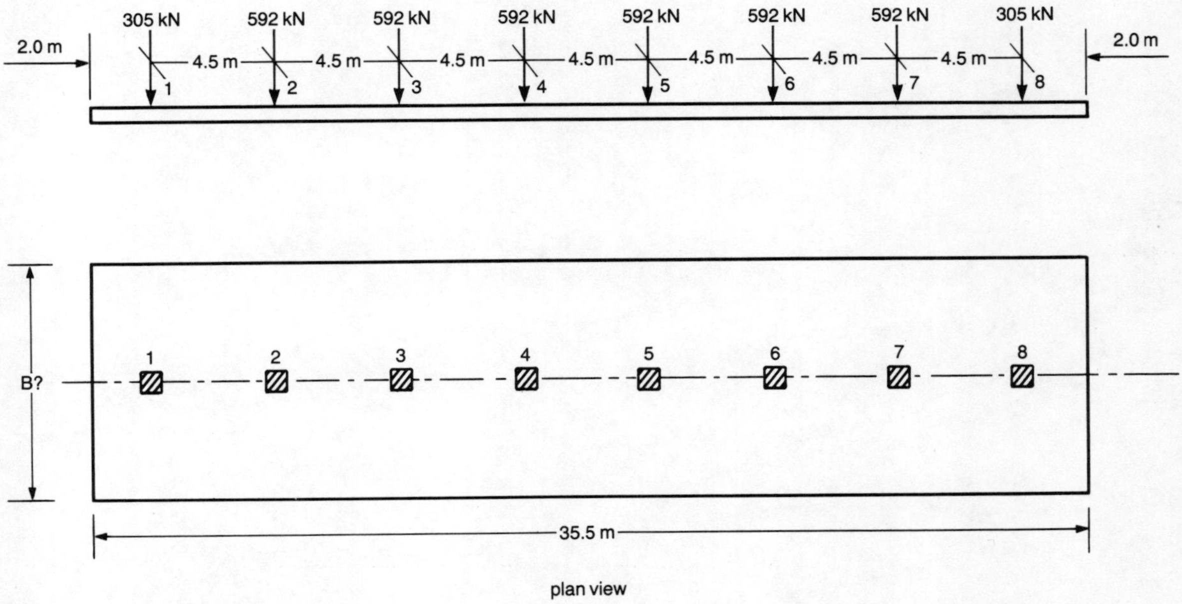

Fig. Q6.6

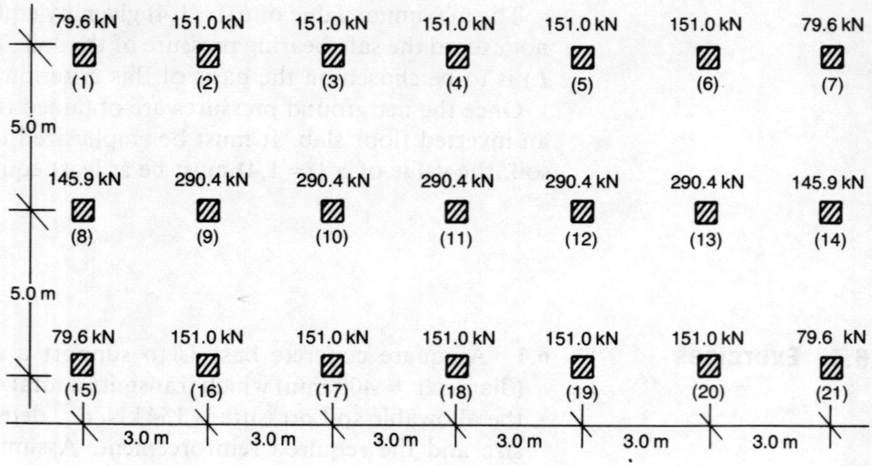

6.6 Design a suitable raft foundation for the columns shown in Fig. Q6.6. The site material consists of soft clay with a safe bearing capacity of only $30\,\text{kN/m}^2$. Assume a 1:2:4 concrete mix and mild steel reinforcement.

6.7 Attempt Exercise 6.6 again, given that the loads on columns (15) to (21) are double the amounts shown in Fig. Q6.6.

7 The Load-factor Method

7.1 Introduction

As stated in chapter 1, in the elastic method of design stress is assumed to be proportional to strain and a constant modular ratio is used. The stress in concrete varies from a maximum value at the extremity of the compression zone to a zero value at the neutral axis [see Fig. 3.1(b)].

However, tests carried out on concrete up to failure have shown that, although the strain varies linearly with distance from the neutral axis, the stress-strain relationship is complex. The complexity is introduced because of the variation in the modulus of elasticity of concrete as failure is approached.

As a result of numerous tests on concrete, it has been found that, provided the maximum strain is limited to $\frac{1}{3}$ per cent, the complex stress distribution in the compression zone can be replaced by a rectangular block (Fig. 7.1). Thus, a constant compressive stress is obtained in the compression zone up to a depth x_0. Generally, $x_0 = 0.85x$, where x is the neutral-axis depth. However, for beams and slabs x_0 can extend to x, provided the value does not exceed $\frac{1}{2}d$.

Tests performed on beams have further shown that, as failure is approached, the compressive stress in concrete can be taken as two-thirds the ultimate cube strength (p_{cu}), while the stress in steel is taken as the yield stress (p_{sy}). The requirements of NCP 1 (clause 4.1.5) are that a general load factor of 1.8 be used.

Fig. 7.1 Stress and strain diagrams for the load-factor method

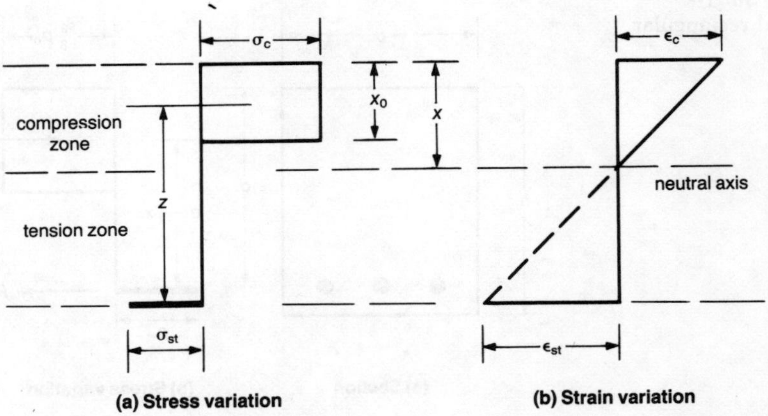

In calculating the ultimate strength, the cube strength (p_{cu}) should be taken as two-thirds of the actual cube strength (p_{ca}) where designed mixes are used, or three-fifths of p_{ca} for nominal mixes.

7.2 Beams

In this section an outline of the design of beams for bending moments, based on the load-factor method, is presented. As usual, analyses are performed on rectangular and simple flanged sections. For more complicated sections, integrations such as those introduced in previous chapters have to be used. Fortunately, because of the constant stress block, analyses of sections based on the load-factor method are generally easier than those based on the modular-ratio technique.

7.2.1 Singly-reinforced rectangular beams

The depth x_0 of the stress block can be taken as equal to the neutral-axis depth x, which in turn is not to exceed half the effective depth d. Therefore, the ultimate compression for a rectangular beam of width b (Fig. 7.2) is:

$$F_{cu} = {}^2/_3\, p_{cu} bx \tag{7.1}$$

while the ultimate tension is:

$$F_{stu} = A_{st}\, p_{sy} \tag{7.2}$$

In addition to the general factor of safety 1.8 that has already been introduced, the use of the factors 5/3 for nominal mixes and 3/2 for designed mixes gives an overall safety factor of 1.8 × 5/3 = 3.0 for nominal mixes, and 3/2 × 1.8 = 2.7 for designed mixes. Therefore, for a rectangular beam, the allowable or 'safe' compression for nominal mixes is ${}^1/_3 F_{cu}$ or:

$$F_c = {}^1/_3({}^2/_3\, p_{cu} bx) = {}^2/_9\, p_{cu} bx \tag{7.3}$$

As the permissible concrete stress p_c is one-third the ultimate cube strength p_{cu} then, for nominal mixes:

$$F_c = {}^2/_9(3p_c)bx = {}^2/_3\, p_c bx \tag{7.4}$$

For designed mixes, the permissible concrete stress p_c is one-third the ultimate

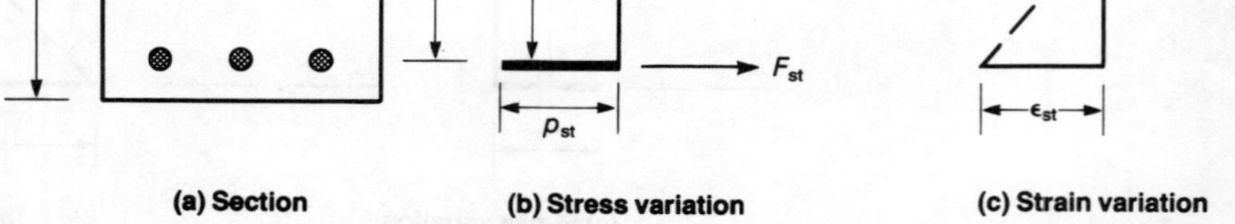

Fig. 7.2 Singly-reinforced rectangular beam

cube strength plus 10 per cent, ie, $p_c = 1.1 \times {}^1\!/_3 p_{cu}$. Therefore, with an overall safety factor of 2.7, the 'safe' compression is:

$$F_c = (1/2.7)F_{cu} = (1/2.7)\,({}^2\!/_3\, p_{cu} bx) \tag{7.5}$$

or, with $p_{cu} = 3p_c/1.1$:

$$F_c = (1/2.7)({}^2\!/_3)\,(3p_c/1.1)bx = {}^2\!/_3\, p_c bx \tag{7.6}$$

which is the same expression as that derived for nominal mixes.

The immediate consequence of equations (7.4) and (7.6) is that the uniform stress σ_c in the compression zone can be replaced by ${}^2\!/_3\, p_c$, as shown in Fig. 7.2(b).

The safe moment of resistance of the section based on concrete is:

$$M_R = F_c z = {}^2\!/_3\, p_c bx(d - x/2) \tag{7.7}$$

Similarly, for steel the ultimate tension is given by equation (7.2) and, with a safety factor of 1.8, the 'safe' tension is:

$$F_{st} = (1/1.8)F_{stu} = (1/1.8)A_{st}\, p_{sy} \tag{7.8}$$

However, the permissible stress in steel p_{st} is 55 per cent of the yield stress p_{sy}, ie, $p_{sy} = p_{st}/0.55$ and:

$$F_{st} = (1/1.8)A_{st}\,(p_{st}/0.55) = A_{st}\, p_{st} \tag{7.9}$$

implying that $\sigma_{st} = p_{st}$, as indicated in Fig. 7.2(b). Therefore, the safe resistance moment based on steel is:

$$M_R = F_{st}\, z = A_{st}\, p_{st}\,(d - x/2) \tag{7.10}$$

For the equilibrium of forces, $F_c = F_{st}$ and:

$$^2\!/_3\, p_c\, bx = A_{st}\, p_{st} \tag{7.11}$$

or

$$x = {}^3\!/_2\, p_{st} A_{st}/bp_c \tag{7.12}$$

As $x \not> {}^1\!/_2 d$, the limiting lever arm is:

$$z = d - {}^1\!/_4 d = {}^3\!/_4 d \tag{7.13}$$

and the limiting moment of resistance based on concrete is:

$$M_R = {}^1\!/_4\, p_c bd^2 = K_c bd^2 \tag{7.14}$$

where $K_c = {}^1\!/_4\, p_c$. It should be recalled that for the modular-ratio method, the limiting value of K_c is given by $K_c = {}^1\!/_2\, p_c n(1 - n/3)$, where n is the neutral-axis factor. When $p_{st} = 126\,\text{N/mm}^2$, $p_c = 7\,\text{N/mm}^2$ and $m = 15$, then the limiting value of $K_c = 1.34\,\text{N/mm}^2$ for the elastic method, whereas $K_c = 1.75\,\text{N/mm}^2$ for the load-factor method, showing that the latter provides a larger moment of resistance when the safe moment is based on concrete alone.

When the safe moment is based on steel the neutral-axis depth x is given by equation (7.12), while the resistance-moment is given by:

$$M_R = A_{st}\, p_{st}(d - {}^3\!/_4\, A_{st}\, p_{st}/bp_c) \tag{7.15}$$

or, with $\rho = A_{st}/bd$

$$K_{st} = M_R/bd^2 = \rho\,(1 - {}^3\!/_4 \rho\, p_{st}/p_c)p_{st} \tag{7.16}$$

which can be compared with the corresponding equation (3.16) for the elastic method.

For a balanced section, the concrete and steel will fail simultaneously and equation (7.11) will yield (with $x = {}^1/_2 d$):

$$A_{st}/bd = \rho = {}^1/_3\, p_c/p_{st} \qquad (7.17)$$

Example 7.1

A reinforced concrete beam is required to carry a bending moment of 58.0kNm.
(a) What depth of concrete is required, assuming a beam width of 280mm?
(b) How many 25mm diameter bars are required for reinforcement?
Assume p_{st} = 126N/mm² and p_c = 7N/mm².

Solution

Assuming a balanced section, equation (7.14) can be used to obtain the effective depth d, as:

$$d = 2\sqrt{(M/p_c b)} = 2\left\{\left(\frac{58.0 \times 10^6}{7 \times 280}\right)\right. = 344\,\text{mm}$$

Using a total cover of 40mm, minimum h = 344 + 40 = 384mm. Therefore, provide an overall depth of 400mm.

From equation (7.17):

$$\rho = A_{st}/bd = {}^1\!/_3\, p_c/p_{st} = {}^1\!/_3(7/126)$$
$$= 0.0185$$
$$A_{st} = 0.0185 \times 280 \times (400 - 40) = 1867\,\text{mm}^2$$

Therefore, use four 25mm ϕ bars (1963mm²).

Comments

This problem is essentially the same as Example 3.1 which was solved using the modular-ratio method. The result required a 440 mm overall depth with three 25 mm diameter bars. It would appear then that the load-factor method tends to require a smaller amount of concrete but a greater proportion of reinforcement than the elastic method, under the same working stresses.

7.2.2 Doubly-reinforced rectangular beams

Consider the doubly-reinforced rectangular beam shown in Fig. 7.3. The force in the compression steel is:

$$f_c = A_{sc}\, p_{sc} \qquad (7.18)$$

where p_{sc} is the permissible compressive stress in steel. The safe compression in the concrete alone is:

$$F_c = {}^2\!/_3\, p_c bx \qquad (7.19)$$

or, with $x = {}^1\!/_2 d$:

$$F_c = {}^1\!/_3\, p_c bd \qquad (7.20)$$

With the total tension $F_{st} = p_{st} A_{st}$ the equilibrium condition $F_{st} = F_c + f_c$ requires that:

$${}^1\!/_3\, p_c bd + A_{sc} p_{sc} = p_{st} A_{st} \qquad (7.21)$$

Fig. 7.3 Doubly-reinforced rectangular section

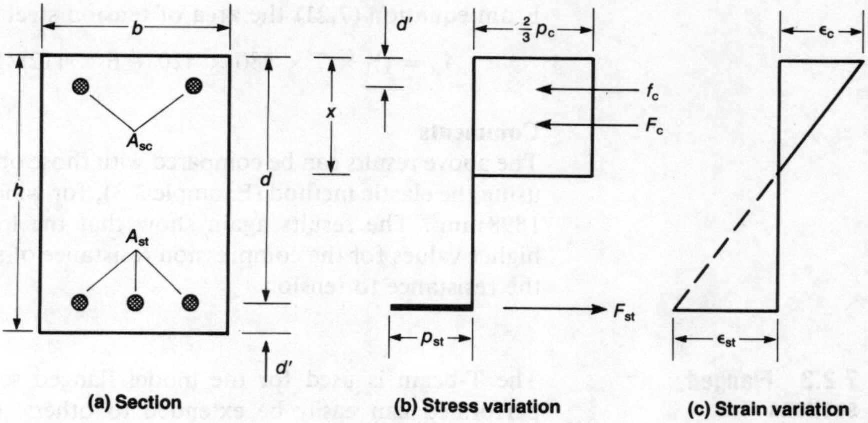

having ignored the area A_{sc} of concrete displaced by the compression reinforcement.

The safe moment of resistance based on concrete alone is:

$$M_R = F_c(\tfrac{3}{4}d) + f_c(d-d') \tag{7.22}$$

or, with equations (7.18) and (7.20):

$$M_R = \tfrac{1}{4}p_c bd^2 + A_{sc}p_{sc}(d-d') \tag{7.23}$$

If the applied moment is M, then to determine the area of compression steel set $M_R = M$ to obtain:

$$A_{sc} = (M - \tfrac{1}{4}p_c bd^2)/p_{sc}(d-d') \tag{7.24}$$

As the moment of resistance of a singly-reinforced rectangular section is $\frac{1}{4}p_c bd^2$ [see equation (7.14)] the term $(M - \frac{1}{4}p_c bd^2)$ simply represents the extra moment carried by the compression steel.

To obtain the balanced reinforcement required, equation (7.24) is introduced into equation (7.21) to give:

$$A_{st} = p_c bd^2(1-4d'/d)/12(d-d')p_{st} + M/(d-d')p_{st} \tag{7.25}$$

Example 7.2

A 280 mm wide rectangular beam is required to carry a bending moment of 86.7 kN m. Given that the depth is restricted to 460 mm, determine the amount of reinforcement required. Assume $p_c = 7$ N/mm^2, $p_{st} = 126$ N/mm^2 and $p_{sc} = 112$ N/mm^2.

Solution

Take $d' = 40$ mm. Therefore, $d = 460 - 40 = 420$ mm.

The moment of resistance M_R of a singly-reinforced section is given by:

$$M_R = \tfrac{1}{4}p_c bd^2$$
$$= \tfrac{1}{4} \times 7 \times 280 \times 420^2 = 86\,436\,000 \text{ N mm}$$

But the applied moment = 86 700 000 N mm. Therefore, compression reinforcement is required.

The extra moment to be carried by the compression steel = 264 000 N mm. Using equation (7.24) the area of compression steel required is given by:

$$A_{sc} = 264\,000/112\,(420-40) = 6 \text{ mm}^2$$

From equation (7.21) the area of tension steel required is given by:

$$A_{st} = (\tfrac{1}{3} \times 7 \times 280 \times 420 + 6 \times 112)/126 = 2183 \text{ mm}^2$$

Comments

The above results can be compared with those obtained for the same problem using the elastic method (Example 3.3), for which $A_{sc} = 698 \text{ mm}^2$ and $A_{st} = 1898 \text{ mm}^2$. The results again show that the load-factor method provides higher values for the compression resistance of sections but lower values for the resistance to tension.

7.2.3 Flanged sections

The T-beam is used for the model flanged section here and the analysis performed can easily be extended to others. Consider the forces on the T-beam shown in Fig. 7.4. The basis for the selection of the flange width b has been outlined in section 3.1.3.

The compression in the flange is:

$$F_c = \tfrac{2}{3} p_c b h_f \tag{7.26}$$

and the contribution from the rib is:

$$f_c = \tfrac{2}{3} p_c b_w (x - h_f) \tag{7.27}$$

The force on the tension steel is $F_{st} = p_{st} A_{st}$. For the equilibrium of forces $F_{st} = F_c + f_c$ and:

$$\tfrac{2}{3} p_c b h_f + \tfrac{2}{3} p_c b_w (x - h_f) = p_{st} A_{st} \tag{7.28}$$

The moment of resistance based on concrete alone is:

$$M_R = F_c(d - \tfrac{1}{2}h_f) + f_c[\tfrac{1}{2}(x - h_f) + (d - x)] \tag{7.29}$$

which, when combined with equations (7.26) and (7.27), gives:

$$M_R = \tfrac{2}{3} p_c [b h_f (d - \tfrac{1}{2}h_f) + \tfrac{1}{2}(x - h_f)^2 b_w] \tag{7.30}$$

Introducing the limiting neutral-axis depth $x = \tfrac{1}{2}d$:

$$M_R = \tfrac{2}{3} p_c [b h_f (d - \tfrac{1}{2}h_f) + \tfrac{1}{2}(\tfrac{1}{2}d - h_f)^2 b_w] \tag{7.31}$$

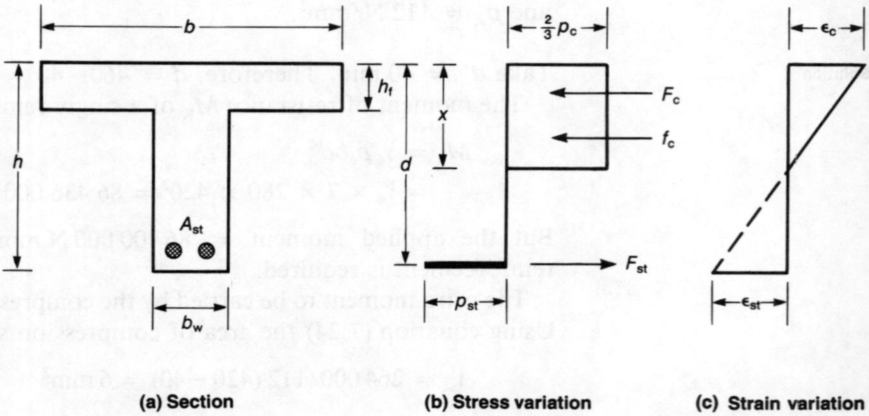

Fig. 7.4 T-section

in which case the lever arm is given by:

$$z = M_R/(F_c + f_c) = \frac{[bh_f(d - {}^1\!/_2 h_f) + {}^1\!/_2({}^1\!/_2 d - h_f)^2 b_w]}{{}^2\!/_3[bh_f + b_w({}^1\!/_2 d - h_f)]} \tag{7.32}$$

If the rib forces are neglected ($f_c << F_c$)

$$z = d - {}^1\!/_2 h_f \tag{7.33}$$

and

$$M_R = {}^2\!/_3 p_c\, bh_f(d - {}^1\!/_2 h_f) \tag{7.34}$$

in which case the moment of resistance based on steel alone is:

$$M_R = p_{st} A_{st}\,(d - {}^1\!/_2 h_f) \tag{7.35}$$

7.3 Solid slabs

The design of solid slabs using the load-factor method is similar to the method described above for beams. A width of 1000 mm is used and the strip is treated as a rectangular beam with the appropriate supporting system.

Example 7.3

A reinforced concrete slab is supported on two parallel block walls which are 3 m apart (centre to centre). If the slab carries a uniformly-distributed load of 10 kN/m², determine the suitable slab thickness and reinforcement. Assume p_c = 7 N/mm² and p_{st} = 126 N/mm².

Solution

The maximum bending moment = $10 \times 10^3\ (3^2)/8$ = 11 250 N m. For a balanced section:

$$K_c = {}^1\!/_4\, p_c = 1.75\ \text{N/mm}^2$$

and

$$M_R = 1.75\, bd^2$$

By setting M_R = 11.25 × 10⁶ N mm and b = 1000 mm, the required value of d is given by:

$$d = \sqrt{\left(\frac{11.25 \times 10^6}{10^3 \times 1.75}\right)} = 80.2\ \text{mm}$$

With a total cover of 24 mm, minimum h = 103.2 mm. But minimum h for deflection (from Table 4.9) is $3 \times 10^3/25$ = 120 mm. Therefore, a slab that is 120 mm thick should be provided.

From equation (7.17) the area of tension steel required is:

$$A_{st} = {}^1\!/_3 \times 7 \times 1000 \times (120 - 24)/126$$
$$= 1778\ \text{mm}^2$$

or, (from Table 4.2), use 16 mm ϕ bars at 100 mm c/c (2011 mm²) bottom.

For distribution, use 10 mm ϕ bars at 175 mm c/c (449 mm²) bottom.

Comments

The above results can be compared with those of Example 4.1, which was an identical problem that was solved using the modular-ratio method and yielded A_{st} = 1061 mm². Again, it can be seen that the load-factor method has required a higher proportion of reinforcement than the elastic method.

7.4 Columns

The treatment of the load-factor method by NCP 1 is actually restricted to beams and slabs. However, the British Code of Practice CP 114 (with which NCP 1 is closely associated) provides guidelines for the design of columns using the load-factor method.

As stated earlier in section 7.1, the compressive stress in concrete is taken as two-thirds of the ultimate cube strength (p_{cu}) acting over a depth $x_0 = 0.85x$. If it is now assumed that the strength of concrete in direct compression is 75 per cent of the actual strength (for designed mixes) or 68 per cent (for nominal mixes) then, with a general load factor of 1.8, the stress in concrete for nominal mixes is:

$$\sigma_{cc} = \frac{0.68}{1.8} \times (^2/_3\, p_{cu}) \tag{7.36}$$

As the permissible stress in direct compression p_{cc} is 76 per cent (NCP 1, clause 3.3.1) of the permissible stress in bending, which in turn is $^1/_3\, p_{cu}$, then:

$$\sigma_{cc} = \frac{0.68}{1.8} \times \frac{2}{3} \times \frac{3p_{cc}}{0.76} \simeq p_{cc} \tag{7.37}$$

Similarly, for designed mixes:

$$\sigma_{cc} = \frac{0.75}{1.8} \times \frac{2}{3} \times \frac{3p_{cc}}{1.1 \times 0.76} \simeq p_{cc} \tag{7.38}$$

where the fact that the permissible (bending) compressive stress is one-third the ultimate cube strength plus ten percent has been introduced.

Hence for columns the safe uniform compressive stress is simply the permissible stress in direct compression (*cf.* beams/slabs where the uniform stress is $^2/_3\, p_c$).

The safe concentric thrust on a concrete column of cross-sectional area A_c (excluding finishes and reinforcing steel) is:

$$N_c = A_c p_{cc} + A_{sc} p_{sc}$$

where A_{sc} is the total area of the reinforcing steel. An identical expression [see equation (5.1)] is used in the modular-ratio method.

In the analysis/design of columns (subjected to combined thrusts and moments) using the load-factor method, three conditions must be identified.

Condition 1: The concrete reaches the limiting strain of $^1/_3$ per cent just as the tension reinforcement reaches the yield point.

This, judging from previous experience, is obviously the 'balanced' condition. With reference to the strain diagram in Fig. 7.5(c) and by the principle of similar triangles:

$$x/(d-x) = \epsilon_c/\epsilon_{st} \tag{7.39}$$

or

$$x/d = \epsilon_c/(\epsilon_{st} + \epsilon_c) \tag{7.40}$$

With $\epsilon_c = {}^1/_3$ per cent $= 1/300$ and $\epsilon_{st} = p_{sy}/E_{st} = 1.8\, p_{st}/E_{st}$:

$$x/d = E_{st}/(E_{st} + 540\, p_{st}) \tag{7.41}$$

Fig. 7.5 Rectangular column subjected to combined thrust and moment

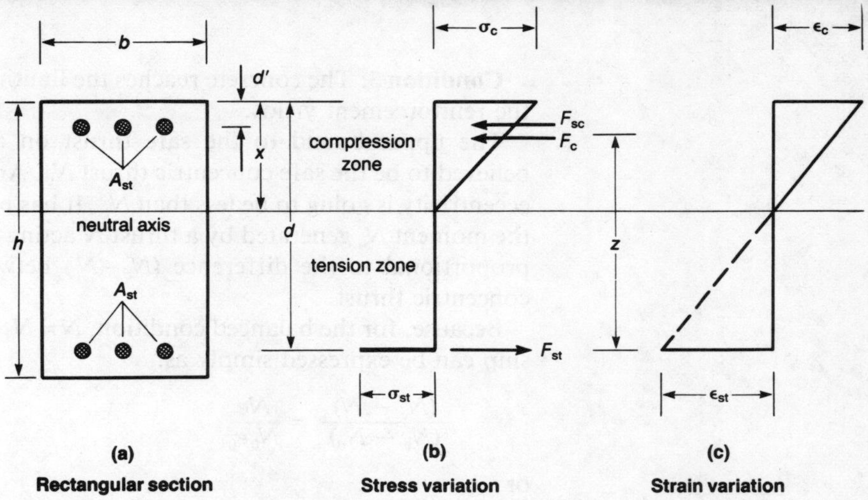

which is an expression for the neutral-axis depth.

The depth of the stress block is:

$$x_0 = 0.85x = 0.85\, E_{st}\, d/(E_{st} + 540\, p_{st}) \tag{7.42}$$

Note that equation (7.42) is not limited to rectangular sections because the derivation has not predetermined the shape of the section.

For the rectangular section in Fig. 7.5 with a thrust N acting at an eccentricity e (from the centroid of section), the equilibrium of forces requires that:

$$N = 0.85xbp_{cc} + A_s{}'p_{sc} - A_s p_{st} \tag{7.43}$$

By taking moments about the location of the tensile reinforcement:

$$\begin{aligned} M &= N(e + \tfrac{1}{2}h - d') \\ &= 0.85\, xbp_{cc}[d - \tfrac{1}{2}(0.85x)] + A_s{}'\, p_{sc}(d - d') \end{aligned} \tag{7.44}$$

For the balanced condition equation (7.41) is simply introduced into equations (7.43) and (7.44), and the balanced thrust N_0 and balanced moment M_0 are obtained as:

$$N_0 = 0.85\Omega_0\, bd\, p_{cc} + A_s{}'\, p_{sc} - A_s\, p_{st} \tag{7.45}$$

$$\begin{aligned} M_0 &= N_0(e_0 + \tfrac{1}{2}h - d') \\ &= 0.85\Omega_0\, bd^2\, p_{cc}[1 - 0.425\Omega_0] + A_s{}'\, p_{sc}(d - d') \end{aligned} \tag{7.46}$$

in which $\Omega_0 = E_{st}/(E_{st} + 540\, p_{st})$ and e_0 is the 'balanced' eccentricity.

Condition 2: The tension reinforcement yields before the concrete reaches the limiting strain.

Under this condition, although equations (7.43) and (7.44) are still valid, equation (7.41) will no longer be appropriate. However, by eliminating x from the two valid equations, the result is a quadratic equation in N given by:

$$\xi_2 N^2 + [\xi_0 + 2\xi_1\xi_2 - d]N + [\xi_2\xi_1{}^2 - \xi_3 - \xi_1 d] = 0 \tag{7.47}$$

where $\xi_0 = e + \tfrac{1}{2}h - d'$; $\xi_1 = A_s p_{st} - A_s{}'\, p_{sc}$; $\xi_2 = \tfrac{1}{2}bp_{cc}$; and $\xi_3 = A_s{}'\, p_{sc}(d - d')$. Equation (7.47) can be solved to obtain the thrust that will lead to the failure of the tensile reinforcement first, in terms of the eccentricity e and the section properties.

Condition 3: The concrete reaches the limiting strain of ⅓ per cent before the reinforcement yields.

The upper bound to the safe thrust on a given column is generally believed to be the safe concentric thrust N_c. Any other safe thrust at a given eccentricity is going to be less than N_c. It has been suggested in CP 114 that the moment N_e generated by a thrust N acting at an eccentricity e is directly proportional to the difference $(N_c - N)$ between the thrust and the safe concentric thrust.

Because, for the balanced condition, $N = N_0$ and $e = e_0$ this linear relationship can be expressed simply as:

$$\frac{(N_c - N)}{(N_c - N_0)} = \frac{Ne}{N_0 e_0}$$

or

$$N = N_c/[1 + (N_c/N_0 - 1)e/e_0] \tag{7.48}$$

which is the appropriate expression for the safe thrust when failure is due to the concrete attaining the limiting strain.

In designing a column for combined thrust and moment the procedure is as follows:

- choose a trial section (eg, for a rectangular column, select b, d, A_s and A_s').
- compute the balanced thrust N_0 and eccentricity e_0 using equations (7.45) and (7.46).
- compare the computed values of N_0 and e_0 with the applied thrust and eccentricity.
- decide, from the comparisons, whether failure will be due to the tension reinforcement (condition 2; $N < N_0$) or due to the concrete reaching the limiting eccentricity (condition 3; $N > N_0$).
- use expression (7.47) or (7.48) to obtain the safe thrusts and see whether the trial section is acceptable.
- iterate until the desired results are obtained.

7.5 Exercises

7.1 Attempt Exercises 3.1 to 3.13 using the load-factor method.

7.2 Redesign the slabs in Exercises 4.1 to 4.3 using the load-factor method.

7.3 Attempt Exercises 5.1 to 5.8 using the load-factor method.

7.4 A singly-reinforced rectangular beam is required to carry a bending moment of M units. Assuming the section is balanced, obtain expressions for the ratios of depths (d_L/d_M) and reinforcement areas (A_L/A_M) where:

d_L = effective depth based on the load-factor method (LFM)
d_M = effective depth based on the modular-ratio method (MRM)
A_L = area of tension reinforcement based on LFM
A_M = area of tension reinforcement based on MRM

Assume p_{st} = permissible stress in steel, p_c = permissible stress in concrete and m = modular ratio.

Part Two: **The Limit State Method**

(designs based on the requirements of CP 110)

8 Introduction to the Limit State Method

8.1 Notation

Some of the notations for the limit state method of design (and thus not used in Part 1 of this book) are given in Table 8.1 overleaf.

8.2 Assumptions

A general introduction to the limit state method of design has been presented earlier in section 1.9.3. The basic philosophy of the limit state design is that the design process must achieve acceptable probabilities such that the structure being designed will not become unfit for the use for which it is required — that is, that the structure will not reach a limit state.

One major difference between the limit state method and the other so-called working-stress methods, such as the modular-ratio method, is the clear separation between the requirements of strength and those of serviceability. Strength requirements seek to make the ultimate load capacity of the structure much greater than the working loads. Serviceability requirements place direct controls on the deflection, cracking and vibration suffered by the structure. Other considerations may include fatigue (for cyclic loads), durability, fire resistance and resistance to lightning.

The ultimate limit state (referring to the strength of the structure at failure) is usually the most critical. Having designed a given structural element to meet the requirements of the ultimate limit state, a check is then performed to ensure that the requirements of the serviceability limit states are indeed satisfied. The processes by which the requirements of the ultimate and serviceability limit states are satisfied are not unconnected because, for example, the choice of appropriate aspect or slenderness ratios (to satisfy serviceability needs) invariably affects the strength of the member. The limit state method also allows for other limit states, for structures designed for unusual or special functions.

In the limit state method, it is assumed that there is a linear strain distribution across the section and a limiting strain of 0.35 per cent at the extremity of the compression zone.

The stress-strain relationships in concrete and steel are as shown in Figs. 8.1 and 8.2, in which γ_m = partial safety factor for load, f_{cu} = characteristic

Table 8.1 Notations for the limit state method

Symbol	Quantity
f_{cu}	characteristic cube strength of concrete
f_y	characteristic strength of reinforcement
N_{uz}	axial load capacity of a column ignoring all bending
Q_R	characteristic imposed load
q	distributed live load
q_k	characteristic live load per unit area
s_b	spacing of bars
s_v	spacing of links along the member
u	perimeter
u_{crit}	length of a critical perimeter
V	shear force due to ultimate loads
V_c	ultimate shear resistance of concrete
V_{c0}	ultimate shear resistance of a section uncracked in flexure
V_{cr}	ultimate shear resistance of a section cracked in flexure
v_c	ultimate shear stress in concrete
v_h	horizontal shear stress per unit area of contact surface
v_t	torsional shear stress
v_{tu}	ultimate torsional shear stress
W_k	characteristic wind load
x_1	shorter dimension of a link
y_0	half the side of an end block
y_{p0}	half the side of a loaded area
y_1	longer dimension of a link
α_c	ratio of the sum of column stiffnesses to the sum of beam stiffnesses
α_{c1}	value of α_c at lower end of column
α_{c2}	value of α_c at upper end of column
α_{cmin}	minimum value of α_{c1} and α_{c2}
α_f	angle of internal friction for column interfaces
α_n	coefficient as a function of column axial loading
α_x, α_y	bending moment coefficients for slabs with no provision to resist torsion at the corners or to prevent the corners lifting
β_b	ratio of beam moments with respect to service stress in beams
β_{cc}	ratio of total creep to elastic deformation
β_{red}	ratio of the reduction in resistance moment
β_x, β_y	bending moment coefficients for slabs with provision to resist torsion and to prevent the corners lifting
β_1	ratio of longer to shorter base sides
γ_f	partial safety factor for load
γ_m	partial safety factor for strength
ϵ_{cs}	shrinkage strain
ϵ_{c1}	strain in concrete at the level of the tendon, at time of loading
ϵ_{c2}	strain in concrete at the centroid of the section, at time of loading
ϵ_{diff}	differential shrinkage strain
ϵ_m	average strain
ϵ_1	strain at the level considered
η	relaxation coefficient
θ_s	angle between compression face and tension reinforcement
μ	coefficient of friction
ξ_s	depth of slab factor
ρ_0	coefficient dependent on percentage of tension and compression steel in the section
ΣA_{sv}	area of shear reinforcement
ΣU_s	sum of the effective perimeters of the tension reinforcement
Δ_{cc}	concrete creep deformation
Δ_{cs}	concrete shrinkage deformation

Fig. 8.1 Short-term stress-strain curve for normal weight concrete (**Note:** f_y in N/mm^2.)

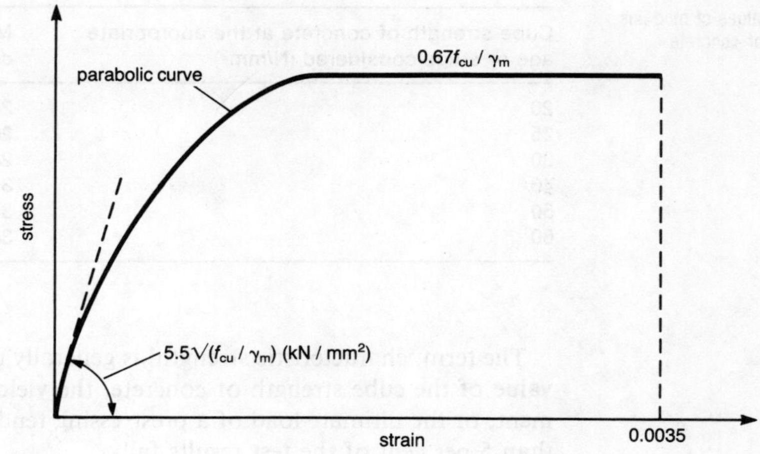

Fig. 8.2 Short-term stress-strain curve for reinforcement (**Note:** f_y in N/mm^2.)

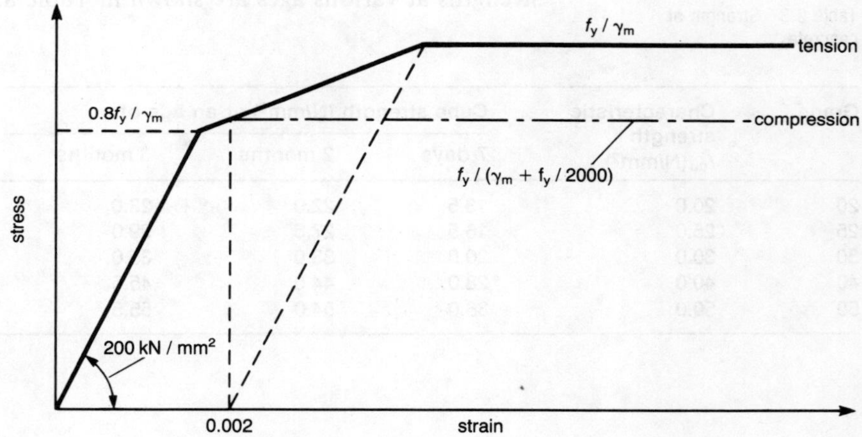

strength for concrete and f_y = characteristic strength for steel.

The British Code of Practice CP 110 allows an alternative assumption of a uniform stress of $0.4f_{cu}$ in concrete throughout the compression zone (*cf.* the load-factor method with a uniform stress of $\frac{2}{3} p_c$ over a depth ranging from $0.85x$ up to a maximum of $0.5d$).

In the absence of better estimates, values of the short-term elastic modulus of concrete (for normal-weight concrete) recommended by CP 110 are presented in Table 8.2. For lightweight aggregate with a density that lies between 1400 kg/m^3 and 2300 kg/m^3 the values given in Table 8.2 should be multiplied by $(D_c/2300)^2$, where D_c is the density of the lightweight aggregate concrete in kg/m^3.

The elastic modulus of steel for all types of loading can be taken as E_s = 200 kN/mm^2. Different values are used for prestressing tendons (see chapter 12).

Table 8.2 Values of modulus of elasticity for concrete

Cube strength of concrete at the appropriate age or stage considered (N/mm^2)	Modulus of elasticity of concrete (kN/mm^2)
20	25
25	26
30	28
40	31
50	34
60	36

The term 'characteristic strength' is generally used in CP 110 to mean the value of the cube strength of concrete, the yield proof stress of reinforcement, or the ultimate load of a prestressing tendon, below which not more than 5 per cent of the test results fall.

The Code classifies concrete into grades, with the grade number being the characteristic cube strength of the concrete. Specified characteristic cube strengths at various ages are shown in Table 8.3.

Table 8.3 Strength of concrete

Grade	Characteristic strength f_{cu} (N/mm^2)	Cube strength (N/mm^2) at an age of				
		7 days	2 months	3 months	6 months	1 year
20	20.0	13.5	22.0	23.0	24.0	25.0
25	25.0	16.5	27.5	29.0	30.0	31.0
30	30.0	20.0	33.0	35.0	36.0	37.0
40	40.0	28.0	44.0	45.5	47.5	50.0
50	50.0	36.0	54.0	55.5	57.5	60.0

Designation	Nominal sizes (mm)	Specified characteristic strength f_y (N/mm^2)
Hot-rolled mild steel (BS 4449)	all sizes	250
Hot-rolled high yield (BS 4449)	all sizes	410
Cold-worked high yield (BS 4461)	up to and including 16	460
	over 16	425
Hand-drawn steel wire	up to and including 12	485

Table 8.4 Strength of reinforcement

The characteristic strengths of reinforcement (which should comply with BS 4449, BS 4461 or BS 4483) are quoted in Table 8.4.

The requirements of CP 110 are that cover to reinforcement should be determined by considerations of fire resistance and durability under the envisaged conditions of exposure. The minimum nominal cover should be the bar diameter. In the case of bundles of three or more bars, the nominal cover should be equal to the size of a single bar of equivalent area.

The recommended values of nominal cover to reinforcement (including links) for dense natural aggregate concrete are shown in Table 8.5.

Condition of exposure	Nominal cover (mm)				
	Concrete grade				
	20	25	30	40	50 and over
Mild: eg, completely protected against weather, or aggressive conditions, except for brief periods of exposure to normal weather conditions during construction.	25	20	15	15	15
Moderate: eg, sheltered from severe rain and against freezing whilst saturated with water. Buried concrete and concrete continuously under water.	—	40	30	25	20
Severe: eg, exposed to driving rain, alternate wetting and drying, and to freezing while wet. Subject to heavy condensation or corrosive fumes.	—	50	40	30	25
Very severe: eg, exposed to sea water or moorland water, and with abrasion.	—	—	—	60	50
Subject to salt used for de-icing:	—	—	50*	40*	25

Table 8.5 Nominal cover to reinforcement

*only applicable if the concrete has entrained air

Table 8.6 shows the recommended lowest grade of concrete for compliance with appropriate use. The grade of concrete required for a specific job depends partly on the nature and the characteristic strength needed to provide the structure with adequate ultimate strength. The grade also depends partly on the exposure conditions and the cover provided to the reinforcement.

Table 8.6 Grades of concrete

Grade	Characteristic strength (N/mm²)	Lowest grade for compliance with appropriate use
7	7.0	plain concrete
10	10.0	
15	15.0	reinforced concrete with lightweight aggregate
20	20.0	reinforced concrete with dense aggregate
25	25.0	
30	30.0	concrete with post-tensioned tendons
40	40.0	concrete with pre-tensioned tendons
50	50.0	
60	60.0	

8.3 Safety factors

The design load and design strength for a given structure, load type and limit state depend on certain partial safety factors. The design load can be obtained by multiplying the characteristic load F_k by the appropriate safety factor γ_f for loads. Thus, design load $= \gamma_f F_k$. The various conditions that the safety factors γ_f take into account have been listed earlier in section 1.9.3.

The design strength is obtained by dividing the characteristic strength f_k

Load combination	dead load (max)	(min)	imposed load (max)	(min)	wind load
1. Dead + imposed load	1.4	1.0	1.6	0.0	0.0
2. Dead + wind load	1.4	0.9	0.0		1.4
3. Dead + imposed + wind load	1.2		1.2		1.2

Table 8.7 Partial safety factor γ_f for loads (ultimate limit state)

by the appropriate partial safety factor γ_m for strength. Thus, design strength $= f_k/\gamma_m$. The safety factors γ_m take into account the possible differences between the strength of the material in the actual structure and the strength derived from test specimens.

The values of the safety factors for the ultimate and serviceability limit states are discussed in the following sections.

8.3.1 Safety factors for the ultimate limit state

For the ultimate limit state, the partial safety factors γ_f for various load combinations and load types are shown in Table 8.7.

When assessing the effect of the design loads on the structure, the arrangement of the loads should be such that they cause the most severe stresses. In some designs it may be necessary to consider the probable effects of excessive loads caused by misuse or accident. In these cases the safety factor γ_f should be taken as 1.05 of the defined loads. Only those loads likely to be acting simultaneously need be considered.

Furthermore, after a structure has sustained localised damage, the γ_f factor should be taken as 1.05 in considering the continued stability of the structure. For this latter case, the loads considered should be those likely to occur before temporary or permanent measures are taken to repair or offset the effect of the damage.

When assessing the strength of a structure, the appropriate values of the partial safety factor γ_m for strength are: 1.5 for concrete and 1.3 for steel.

The Code provides an exception for those special cases where the effects of excessive loads or localised damaged are being considered. In these cases the values of γ_m are: 1.3 for concrete and 1.0 for steel.

8.3.2 Safety factors for the serviceability limit state

The partial safety factors γ_f required for the serviceability limit state, for the indicated load combinations and load types, are presented in Table 8.8.

Again, in assessing the deflection of a structure, the arrangement of the imposed load should be that which causes the largest deflection. The design loads based on the γ_f values in Table 8.8 apply only when estimating the immediate deflections of a structure. In most other cases, it may be necessary to estimate also the additional time-dependent deflections due to creep, shrinkage and temperature.

Table 8.8 Partial safety factors γ_f for loads (serviceability limit state)

Load combination	dead load	imposed load	wind load
1. Dead + imposed load	1.0	1.0	0.0
2. Dead + wind load	1.0	0.0	1.0
3. Dead + imposed + wind load	1.0	0.8	0.8

When assessing the deflections of a structure, the appropriate value of the safety factor γ_m is 1.0, for both concrete and steel. For this purpose, the properties of the materials that are relevant to deflection assessment (eg, moduli of elasticity, creep, shrinkage, etc) should be taken as those associated with the characteristic strengths of the materials. For the purpose of assessing the stresses or crack widths at any section within a structure, the values of γ_m should be taken as: 1.3 for concrete and 1.0 for steel.

Note that the prescribed stress-strain curves for concrete and reinforcement shown in Figs. 8.1 and 8.2 depend on the partial safety factor γ_m. The use of these curves for the analysis and/or design of reinforced concrete sections in beams, slabs and columns is outlined in chapters 9 to 11.

8.4 Analysis

In analysing a structure for the various limit states CP 110 allows the use of any method that provides as accurate a representation of the behaviour of the structure as is practicable. However, certain methods and assumptions are put forward in the Code (clauses 2.4.3, 2.4.4 and 3.2) and these are outlined below.

Generally, when analysing a structure, to determine the force distributions within the structure the material properties assumed may be those associated with the characteristic strengths, regardless of which limit state is being considered.

On the other hand, when any section within a structure is being analysed, the material properties should be assumed to be those associated with the design strengths appropriate to the particular limit state under consideration. (It is hoped here that the student really appreciates the basic difference between the analysis of a *structure*, or an element of it, and that of a *cross-section of the structure.*)

When the force distribution within a structure is obtained on the basis of an elastic analysis, the relative stiffnesses of the members may be based on any of the following:

- the **concrete section**, defined as the entire cross-section, ignoring the reinforcement;
- the **gross section**, defined as the entire concrete cross-section, including the reinforcement, on the basis of modular ratio;
- the **transformed section**, defined as the compression area of the concrete cross-section combined with the reinforcement, on the basis of modular ratio.

Similarly, when considering serviceability limit states, and when elastic analysis is used to determine the force distribution throughout the structure, the relative stiffnesses may be based on any of the preceding three considerations.

However, in the analysis of a section, the Code requires that the strength be assessed by inelastic analysis based on the short-term stress-strain curves (such as those shown in Figs. 8.1 and 8.2).

Furthermore, in determining the deflections of a structure, the curvature of any section should be computed taking into account fully the influence of shrinkage and creep.

The Code also allows the use of suitable models for the purpose of designing and/or analysing the prototype, provided the work is carried out by engineers with relevant experience using suitable equipment.

The recommendations of CP 110 (clause 3.2) in respect of the analysis of framed structures are outlined below.

If, in a single-span continuous frame:

- the ratio of beam span to supporting column height exceeds 3; and
- the reinforcement in the beam is less than 1 per cent;

then either:

- the cracked EI (I = second moment of area) value for the beam should be used in calculating the distribution of moments in the frame;

or:

- the bending moment transmitted to the column should be limited by ensuring that the top reinforcement in the beam at the support under overload will yield before the concrete on the inside of the column attains ultimate stress.

When a framed structure supports vertical loads only, the moments, loads and shear forces used in the design of the elements (ie, monolithic beams and columns) may be based on an elastic analysis of a series of sub-frames. The loading arrangement may consist of:

- alternate spans loaded with total ultimate load ($1.4G_k + 1.6Q_k$), and all other spans loaded with minimum dead load ($1.0G_k$), where G_k = the characteristic dead load and Q_k = the characteristic imposed load;
- any two adjacent spans loaded with total ultimate load ($1.4G_k + 1.6Q_k$), and all other spans loaded with minimum dead load ($1.0G_k$).

The alternative to the preceding is for the moments and forces in individual beams to be found by considering a simplified sub-frame, consisting only of the beam (with the columns attached to the ends of the beam) and the beams, if any, on either side. In this case the column, and the beam ends remote from the particular beam under consideration, should be assumed to be fixed, and the outer beams to possess only half their actual stiffness. The loading arrangement here may consist of:

- the centre beam loaded with total ultimate load ($1.4G_k + 1.6Q_k$), and the other two loaded with minimum dead load ($1.0G_k$);
- any two of the three beams loaded with total ultimate load ($1.4G_k + 1.6Q_k$), and the other beam loaded with minimum dead load ($1.0G_k$).

For framed structures supporting both vertical and lateral loads, it may be necessary to account for sway effects if the columns are slender (ie, ratio of effective length to least lateral dimension exceeds 12). However, in most cases, the design may be based on the various considerations outlined above for frames carrying vertical loads only.

If, in particular, the effects of the lateral loads are such as to make the analysis simpler by considering vertical loads only, then the moments, loads and shear forces should be obtained from the sum of the corresponding values derived from:

- an elastic analysis of a series of sub-frames, each consisting of the beams at one level, together with the columns above and below assumed to be fixed at the ends that are remote from those beams. Lateral loads should be ignored and all beams considered to be loaded with their full design load ($1.2G_k + 1.2Q_k$);
- an elastic analysis of the complete frame, assuming points of contraflexure at the centres of all beams and columns, ignoring dead and

imposed loads, and considering only the design wind load ($1.2W_k$) on the structure (where W_k is the characteristic wind load).

In certain cases, it may be necessary to consider the load combination (2) in Table 8.7. That is, the design load = $(0.9G_k + 1.4W_k)$, which combines the minimum dead load with the recommended design load due to wind.

Under certain specified conditions, CP 110 allows the redistribution of moments (clause 3.2.2.3). These conditions are:

- equilibrium between internal forces and external loads must be maintained under each appropriate combination of ultimate loads.
- the ultimate resistance moment provided at any section of a member must not be less than 70 per cent of the moment at that section obtained from an elastic maximum-moments diagram covering all appropriate combinations of ultimate loads.
- the elastic moment at any section in a member due to a particular combination of ultimate loads should not be reduced by more than 30 per cent of the numerically-largest moment given anywhere by the elastic maximum-moments diagram for that particular member, covering all combinations of ultimate loads. The percentage reduction is limited to 10 in structures exceeding four storeys, and in which the structural frame provides the lateral stability.
- where, as a result of redistribution, the ultimate resistance moment at a section is reduced, the neutral axis depth x of the section resisting the reduced moment should not be greater than:

$$x = (0.6 - \beta_{red})d \tag{8.1}$$

in which d is the effective depth and β_{red} is the ratio of the reduction in resistance moment to the numerically-largest moment given anywhere by the elastic maximum-moments diagram for that particular member, covering all appropriate combinations of ultimate loads.

9 Beams

9.1 Ultimate limit state

Although all the limit states should be considered in a given design, normal practice is to design on the basis of the most critical limit state and then to check that the remaining limit states will not be reached. For reinforced concrete structures, it is usual to assume that the ultimate limit state is the most critical. The design methods thus generally use the partial safety factors appropriate for the ultimate limit state, followed by the pertinent recommendations that ensure that the serviceability limit states (especially for deflection and cracking) are not reached.

As stated in sections 1.9.3, 8.2 and 8.3, when analysing a given cross-section for the purpose of determining the ultimate moment of resistance, it is assumed that:

- sections that are plane before bending remain plane after bending. This implies that the strain in both concrete and reinforcement varies directly with the distance from the neutral axis;
- the tensile strength of the concrete is negligible;

Fig. 9.1 Stress and strain distribution in concrete at ultimate limit state

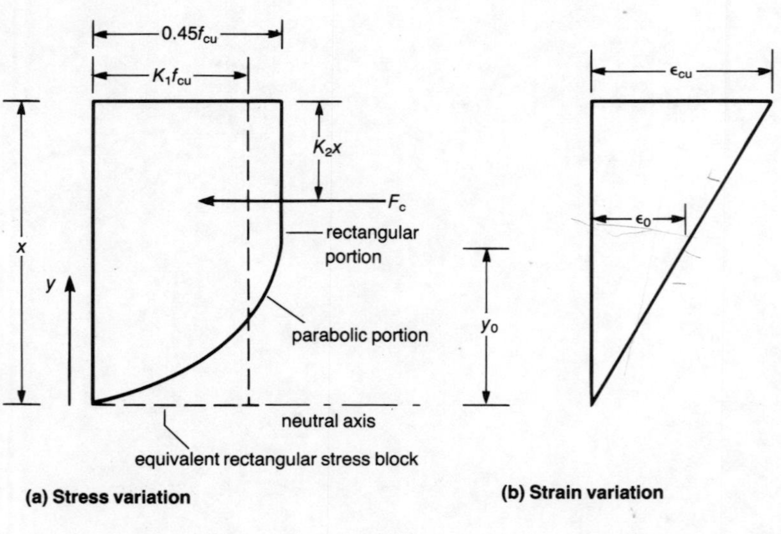

- the stresses in concrete can either be based on the stress-strain curve of Fig. 8.1 (with $\gamma_m = 1.5$) or be taken as being equal to $0.4f_{cu}$ over the entire compression zone. In both cases the maximum strain in concrete at failure is taken as $\epsilon_{cu} = 0.35$ per cent. For singly-reinforced beams the depth of concrete in compression is limited to half the effective depth of the beam (*cf.* the load-factor requirement);
- the stresses in the reinforcement are based on the stress-strain curves of Fig. 8.2 (with $\gamma_m = 1.15$).

CP 110 provides certain design formulae (part 1) and relevant charts (parts 2 and 3) for the most common sections. However, in the same way that the working stress methods have been developed, these formulae will (as much as possible) be derived here from first principles by using the prescribed stress-strain curves and the other relevant assumptions listed above.

To this end, consider Fig. 9.1 which shows the recommended stress and strain variations in the concrete for any given structural element. Let the rectangular-parabolic curve be replaced by an 'equivalent' rectangular block (dashed line) of width $K_1 f_{cu}$, depth x and whose centroid is at a distance $K_2 x$ from the top of the compression zone.

The strain ϵ_0 in concrete at the point (distance y_0 from the neutral axis) where the parabolic portion of the stress curve meets the rectangular block, is related to the characteristic strength f_{cu} by the expression:

$$\epsilon_0 = \sqrt{f_{cu}}/5000 \qquad (9.1)$$

with f_{cu} in kN/mm^2.

The rectangular-parabolic stress distribution is given by the expressions:

$$\sigma_c = 0.45 f_{cu}\,(y/y_0)(2 - y/y_0) \qquad 0 \le y \le y_0 \qquad (9.2)$$

$$\sigma_c = 0.45 f_{cu} \qquad y_0 \le y \le x \qquad (9.3)$$

in which y is the distance measured from the neutral axis (see Fig. 9.1).

The area of the rectangular-parabolic stress block is:

$$A = {}_0\!\int^x \sigma_c \mathrm{d}y = 0.45 f_{cu}\,[\,{}_0\!\int^{y_0} y/y_0(2 - y/y_0)\mathrm{d}y + {}_{y_0}\!\int^x \mathrm{d}y]$$

$$= 0.45 f_{cu}(x - y_0/3) \qquad (9.4)$$

Now, from the strain diagram and using similar triangles:

$$y_0 = x\epsilon_0/\epsilon_{cu} \qquad (9.5)$$

That is:

$$A = 0.45 f_{cu}\, x(1 - {}^1\!/_3\epsilon_0/\epsilon_{cu}) \qquad (9.6)$$

As the area of the equivalent stress block is $xK_1 f_{cu}$, set:

$$A = xK_1 f_{cu} = 0.45 f_{cu}\, x(1 - {}^1\!/_3\epsilon_0/\epsilon_{cu})$$

or

$$K_1 = 0.45(1 - {}^1\!/_3\epsilon_0/\epsilon_{cu}) \qquad (9.7)$$

By using equation (9.1) and setting $\epsilon_{cu} = 0.0035$:

$$K_1 = 0.45(1 - \sqrt{f_{cu}}/52.5) \qquad (9.8)$$

If the commonly-used characteristic concrete strengths varying between

20 and 60 N/mm² are considered, the average value K_1 can be obtained as:

$$K_1 = \frac{0.45}{(60-20)} {}_{20}\int^{60}(1 - \sqrt{f_{cu}}/52.5)\mathrm{d}f_{cu}$$

or

$$\bar{K}_1 \simeq 0.4 \qquad (9.9)$$

which is the basis for the recommended (alternative) uniform stress $0.4f_{cu}$ in concrete.

To locate the centroid of the stress block take area moments about the neutral axis. Thus:

$$(xK_1f_{cu})(x - xK_2) = {}_0\int^x \sigma_c y\mathrm{d}y$$
$$= 0.45f_{cu}[{}_0\int^{y_0}(y^2/y_0)(2 - y/y_0)\mathrm{d}y + {}_{y_0}\int^x y\mathrm{d}y] \qquad (9.10)$$

Putting $y_0 = x\epsilon_0/\epsilon_{cu}$, $\epsilon_0 = \sqrt{f_{cu}}/5000$ and $\epsilon_{cu} = 0.0035$, and with some algebra, it is possible to obtain:

$$K_2 = \frac{2 + (2 - \sqrt{f_{cu}}/17.5)^2}{12(1 - \sqrt{f_{cu}}/52.5)} \qquad (9.11)$$

In the simplified stress block given in CP 110, $K_1 = 0.4$ and $K_2 = 0.5$. With the values of K_1 and K_2 given in equations (9.8) and (9.11) respectively, the ultimate moment of resistance of various reinforced concrete sections may now be derived.

9.1.1 Singly-reinforced rectangular sections

Fig. 9.2 shows the rectangular section of a reinforced concrete beam and the stress in the reinforcement is given by:

Note that the stress σ_{st} in steel is related to the strain ϵ_{st} by the tri-linear curve of Fig. 8.2, which is reproduced in Fig. 9.3 for tension reinforcement. If the strains at the 'kinks' on the stress-strain curve are ϵ_1 and ϵ_2, then the stress in the reinforcement is given by:

$$\sigma_{st} = 200\,000\,\epsilon_{st}; \quad 0 \leq \epsilon_{st} \leq \epsilon_1$$
$$= [4 + (\epsilon_{st} - \epsilon_1)/(\epsilon_2 - \epsilon_1)]f_y/5.75; \quad \epsilon_1 \leq \epsilon_{st} \leq \epsilon_2$$
$$= f_y/1.15 = 0.87f_y; \quad \epsilon_{st} \geq \epsilon_2 \qquad (9.12)$$

in which f_y is the yield stress in the reinforcement and (with $\gamma_m = 1.15$) the strains ϵ_1, ϵ_2 are given by:

$$\epsilon_1 = \frac{0.8f_y}{1.15 \times 200\,000} = f_y/287\,500 \qquad (9.13)$$

and

$$\epsilon_2 = 0.002 + \frac{f_y}{1.15 \times 200\,000} = 0.002 + f_y/230\,000 \qquad (9.14)$$

From the strain variation across the section [Fig. 9.2(c)]:

$$x/(d-x) = \epsilon_{cu}/\epsilon_{st}$$

or

$$\epsilon_{st} = (d/x - 1)\epsilon_{cu} \qquad (9.15)$$

Fig. 9.2 Singly-reinforced rectangular section

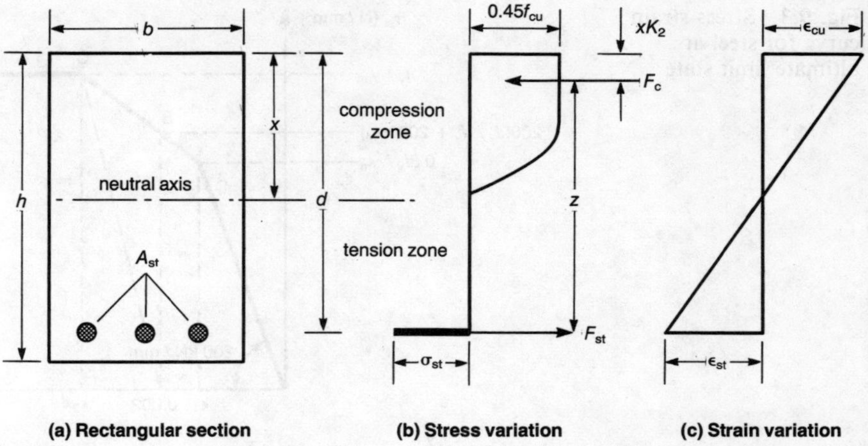

which provides the relationship for the stress in steel as a function of the neutral axis factor x/d. The compression in concrete is:

$$F_c = K_1 f_{cu} xb \tag{9.16}$$

while the tension in steel is:

$$F_{st} = \sigma_{st} A_{st} \tag{9.17}$$

where σ_{st} is given by equation (9.12).

For equilibrium, $F_c = F_{st}$ or:

$$K_1 f_{cu} xb = \sigma_{st} A_{st} \tag{9.18}$$

If the steel has yielded, then $\sigma_{st} = 0.87 f_y$ [see equation (9.12)] and:

$$x = 0.87 f_y A_{st} / K_1 b f_{cu} \tag{9.19}$$

or, with $\rho = A_{st}/bd$:

$$x/d = 0.87 f_y \rho / K_1 f_{cu} \tag{9.20}$$

The lever arm is given by $z = d - K_2 x$ or:

$$z/d = 1 - 0.87 f_y K_2 \rho / K_1 f_{cu} \tag{9.21}$$

The ultimate moment of resistance based on steel is:

$$M_u = 0.87 f_y A_{st} z \tag{9.22}$$

and the ultimate moment of resistance based on concrete is:

$$M_u = K_1 f_{cu} xbz \tag{9.23}$$

For the simplified stress-block, $K_1 = 0.4$, $K_2 = 0.5$ and the ultimate resistance moment based on steel is:

$$M_u = 0.87 f_y A_{st} (1 - 1.1 f_y \rho / f_{cu}) \tag{9.24}$$

and the ultimate moment of resistance based on concrete (with $x/d = 0.5$, $z/d = 0.75$) is:

$$M_u = 0.15 f_{cu} bd^2 \tag{9.25}$$

Fig. 9.3 Stress-strain curve for steel at ultimate limit state

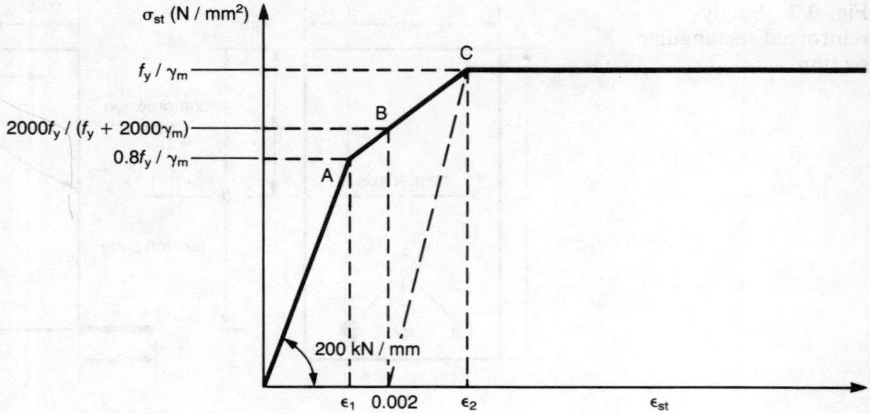

For a balanced section, the concrete and steel will fail simultaneously and $\epsilon_{st} = \epsilon_2$, the strain in the reinforcement at yield point [see Fig. (9.3)]. By combining equations (9.15) and (9.20) with $\epsilon_{st} = \epsilon_2$, the 'balanced' steel proportion is given by:

$$\rho_0 = 1.15K_1 f_{cu}/(1 + \epsilon_2/\epsilon_{cu})f_y \tag{9.26}$$

in which $\epsilon_{cu} = 0.0035$ and ϵ_2 is given by equation (9.14).

If $\rho > \rho_0$, the section is over-reinforced (the concrete will fail first) whereas if $\rho > \rho_0$ the section is under-reinforced (the steel will yield first). As stated earlier in chapter 1, an over-reinforced section is undesirable because failure occurs without ample warning.

If the section is over-reinforced, set $\sigma_{st} = \epsilon_{st} E_{st}$ in equation (9.18) and then eliminate E_{st} using equation (9.15) to obtain:

$$\left(\frac{K_1 f_{cu}}{\rho E_{st}}\right)\left(\frac{x}{d}\right)^2 + \epsilon_{cu}\left(\frac{x}{d}\right) - \epsilon_{cu} = 0 \tag{9.27}$$

which is a quadratic equation for the neutral axis for an over-reinforced section. After rejecting the unrealistic negative root, the neutral axis factor is given by:

$$x/d = \Omega_0[\sqrt{(1 + 2/\Omega_0)} - 1] \tag{9.28}$$

in which $\Omega_0 = \epsilon_{cu}\,\rho E_{st}/2K_1 f_{cu}$. Note that the elastic modulus E_{st} depends on the strain ϵ_{st} in steel (Fig. 9.3) and in turn the strain depends on the neutral axis factor [equation (9.15)]!

Now that all the relevant equations have been derived, the basic problem is how to use them for actual design/analysis of beam sections. In the analysis of a section the pertinent data relating to the section size (b, d), and reinforcement area (A_{st}) will normally be known. The problem involves obtaining the stress (or strain) in the steel and the ultimate moment of resistance of the given section. The location of the position of the neutral axis is the first major hurdle and by virtue of the complex equations involved (at least when the general rectangular-parabolic stress block is used), the process may be iterative.

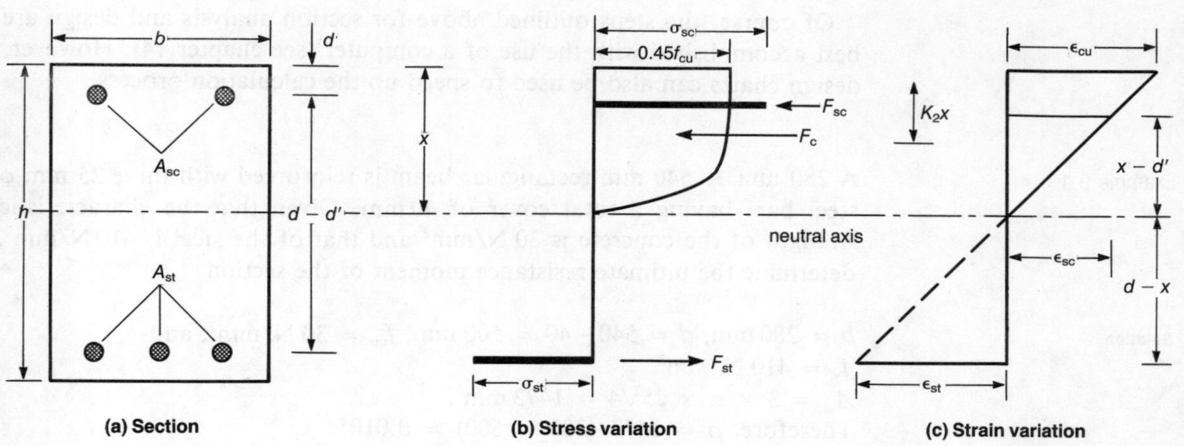

Fig. 9.4 Doubly-reinforced rectangular section

The various steps involved in the analysis of a section are as follows:

- compute the steel proportion $\rho = A_{st}/bd$ and the balanced amount ρ_0 [equation (9.26)];
- if $\rho < \rho_0$ the section is under-reinforced. Obtain ultimate resistance moment using equations (9.21) and (9.22), or alternatively use equation (9.24) if the simplified stress-block is preferred;
- if $\rho > \rho_0$ the section is over-reinforced. Compute the steel proportion ρ_1 when the strain in the steel is ϵ_1 (see Fig. 9.3) by computing x/d from equation (9.15) and setting $\sigma_{st} = 0.7f_y$ in equation (9.18). If $\rho > \rho_1$ then compute x/d from equation (9.28) with $E_{st} = 200\,\text{kN/mm}^2$. Otherwise (ie, $\rho < \rho_1$) compute x/d using equation (9.28) with E_{st} obtained from the slope of the line AB in Fig. 9.3. Then determine the ultimate resistance moment using equation (9.23), with $z = d - K_2x$.

In a typical design problem the appropriate section size (eg, d for prescribed b) and/or reinforcement amount will be required for a given ultimate moment of resistance M_u. The usual steps (when the design is based on the rectangular-parabolic concrete stress block) are as follows:

- compute the balanced steel proportion ρ_0 from equatioin (9.26);
- choose a convenient steel proportion such that $\rho < \rho_0$, so that the section is under-reinforced;
- obtain the neutral-axis factor x/d using equation (9.19);
- obtain the lever-arm factor from $z/d = 1 - K_2(x/d)$;
- compute the resistance moment coefficient $K_c = M_u/bd^2$ using equation (9.23), as $K_c = K_1 f_{cu}(x/d)(z/d)$;
- obtain a suitable effective depth from $d = \sqrt{M_u/K_c b}$;
- compute the area of steel from $A_{st} = \rho\, bd$.

Note that CP 110 limits the neutral-axis factor to 0.5 for singly-reinforced beams. This condition must always be met.

The design process is much easier when the simplified stress curve is used. The steps are simply to:

- obtain the effective depth d from equation (9.25); and
- compute the area A_{st} of steel required from equation (9.18).

The simplified formulae [equations (9.24) and (9.25)] are generally not suitable for the analysis of sections.

Of course, the steps outlined above for section analysis and design are best accomplished with the use of a computer (see chapter 14). However, design charts can also be used to speed up the calculation process.

Example 9.1

A 280 mm × 540 mm rectangular beam is reinforced with three 25 mm ϕ steel bars having a total cover of 40 mm. Given that the characteristic strength of the concrete is 30 N/mm² and that of the steel is 410 N/mm², determine the ultimate resistance moment of the section.

Solution

$b = 280$ mm, $d = 540 - 40 = 500$ mm, $f_{cu} = 30$ N/mm², and $f_y = 410$ N/mm².
$A_{st} = 3 \times \pi \times 25^2/4 = 1473$ mm².
Therefore, $\rho = 1473/(280 \times 500) = 0.0105$.
From equations (9.7) and (9.11):

$$K_1 = 0.45(1 - \sqrt{30}/52.5) = 0.403$$

$$K_2 = \frac{2 + (2 - \sqrt{30}/17.5)^2}{12(1 - \sqrt{30}/52.5)} = 0.417$$

Using equation (9.14) $\epsilon_2 = 0.002 + 410/230\,000 = 3.783 \times 10^{-3}$
Therefore, $\epsilon_2/\epsilon_{cu} = 1.08$
Using equation (9.26):

$$\rho_0 = \frac{1.15 \times 0.403 \times 30}{(1 + 1.08) \times 410} = 0.016$$

Therefore, $\rho < \rho_0$, indicating that the beam is under-reinforced.
The lever arm is given by equation (9.21) as:

$$z = \left(1 - \frac{0.87 \times 410 \times 0.0105}{0.403 \times 30}\right) \times 500 = 435 \text{ mm}$$

and the ultimate resistance moment, from equation (9.22), is:

$$M_u = 0.87 \times 410 \times 1473 \times 435 \times 10^{-6}$$
$$= 229 \text{ kN m.}$$

Example 9.2

Determine the depth d and the amount of reinforcement required for a 280 mm wide rectangular beam, given that the ultimate moment of resistance is 240 kN m. Assume $f_{cu} = 30$ N/mm² and $f_y = 250$ N/mm².

Solution

From Example 9.1, for $f_{cu} = 30$ N/mm²; $K_1 = 0.403$; and $K_2 = 0.417$:

$$\epsilon_2 = 0.002 + 250/230\,000 = 3.087 \times 10^{-3}$$

$$\epsilon_2/\epsilon_{cu} = 0.882$$

$$\text{Therefore, } \rho_0 = \frac{1.15 \times 0.403 \times 30}{(1 + 0.882) \times 250} = 0.03$$

Choose $\rho = 0.025$, say, so that the section is under-reinforced.
Using equation (9.19):

$$x/d = \frac{0.87 \times 250 \times 0.025}{0.403 \times 30} = 0.45$$

(which is less than 0.5). Therefore:

$$z/d = 1 - K_2(x/d) = 1 - 0.417 \times 0.45 = 0.812$$
$$K_c = 0.403 \times 30 \times 0.45 \times 0.812 = 4.42\ \text{N/mm}^2$$

Therefore, the required value of d is given by:

$$d = \sqrt{\left(\frac{240 \times 10^6}{280 \times 4.42}\right)} = 440\ \text{mm}$$

and the amount of reinforcement required is:

$$A_{st} = 0.025 \times 280 \times 440 = 3080\ \text{mm}^2$$

If the simplified stress block is used, then from equation (9.25) the required value of d is given by:

$$d = \sqrt{\left(\frac{240 \times 10}{0.15 \times 30 \times 280}\right)} = 436\ \text{mm}$$

and from equation (9.18), with $x = \frac{1}{2}d$ and $K_1 = 0.4$:

$$A_{st} = \frac{0.4 \times 30 \times 436 \times 280}{0.87 \times 250 \times 2} = 3368\ \text{mm}^2$$

9.1.2 Doubly-reinforced rectangular sections

Consider the doubly-reinforced rectangular section shown in Fig. 9.4. The force in the tension reinforcement is $F_{st} = \sigma_{st}A_{st}$, while that in the compression steel is $F_{sc} = \sigma_{sc}A_{sc}$. As the compression in concrete (ignoring the area A_{sc} displaced by the steel) is $F_c = K_1 f_{cu} xb$, the equilibrium condition $F_c + F_{sc} = F_{st}$ requires that:

$$K_1 f_{cu} xb = \sigma_{st}A_{st} - \sigma_{sc}A_{sc} \qquad (9.29)$$

or, dividing through by bd:

$$K_1 f_{cu}(x/d) = \rho\sigma_{st} - \rho'\sigma_{sc} \qquad (9.30)$$

in which $\rho = A_{st}/bd$ and $\rho' = A_{sc}/bd$.

From the strain diagram the strains ϵ_{st} in the tension steel and ϵ_{sc} in the compression steel are given by:

$$\epsilon_{st} = (d - x)\epsilon_{cu}/x \qquad (9.31)$$

$$\epsilon_{sc} = (x - d')\epsilon_{cu}/x \qquad (9.32)$$

where $\epsilon_{cu} = 0.0035$.

By taking moments about the location of the tension reinforcement, the ultimate resistance moment is given by:

$$M_u = F_{sc}(d - d') + F_c(d - K_2x) \qquad (9.33)$$

or, with $F_{sc} = \sigma_{sc}A_{sc}$ and $F_c = \sigma_{st}A_{st} - \sigma_{sc}A_{sc}$:

$$M_u = A_{sc}\sigma_{sc}(K_2x - d') + \sigma_{st}A_{st}(d - K_2x) \qquad (9.34)$$

Thus:

$$M_u/bd^2 = \rho'\sigma_{sc}[K_2(x/d) - (d'/d)] + \rho\sigma_{st}[1 - K_2(x/d)] \qquad (9.35)$$

If the simplified rectangular stress block is used then $K_1 = 0.4$, $K_2 = 0.5$

and to obtain the ultimate resistance moment set (see Fig. 9.3) $\sigma_{st} = 0.87f_y$, $\sigma_{sc} = 0.72f_y[\simeq 2000f_y/(f_y + 2000\gamma_m); \gamma_m = 1.15]$, $x/d = 0.5$, so that $F_c = 0.4f_{cu}(^1/_2 d)b = 0.2f_{cu}bd$, $z = d - K_2x = 0.75d$, $F_{sc} = 0.72f_y A_{sc}$, and with these in equation (9.33):

$$M_u = 0.72f_yA_{sc}(d-d') + 0.15f_{cu}bd^2 \tag{9.36}$$

while equation (9.30) becomes:

$$0.87f_yA_{st} = 0.2f_ubd + 0.72f_yA_{sc} \tag{9.37}$$

Because, for the simplified stress block, x/d is limited to 0.5, the limiting strain in the compression steel is given by equation (9.32) as:

$$\epsilon_{sc}\text{ (limiting)} = (1-2d'/d)\epsilon_{cu} \tag{9.38}$$

From Fig. (8.2), if the compression reinforcement is to yield, ϵ_{sc} (limiting) ≥ 0.002, that is:

$$(1-2d'/d)\epsilon_{cu} \geq 0.002$$

or, with $\epsilon_{cu} = 0.0035$:

$$d'/d \leq 0.21 \tag{9.39}$$

which is why CP 110 has limited d'/d to 0.2 as the condition for using the formulae given by equations (9.36) and (9.37).

When using the rectangular-parabolic stress block, CP 110 places no restrictions on the value of the neutral-axis factor for doubly-reinforced beams for which, therefore, x/d may exceed 0.5.

Given all the properties of a given doubly-reinforced section, the steps involved in the determination of the ultimate resistance moment are:

- guess a trial neutral axis factor, x/d;
- compute the strains in both compression and tension reinforcement, using equations (9.31) and (9.32);
- determine the stresses in both compression and tension reinforcement by using the stress-strain curve of Fig. 8.2 (or Fig. 9.3);
- compute both lhs and rhs of the equilibrium equation (9.30) and check if it is satisfied (the chances are that it is not);
- if lhs < rhs, then trial x/d is too small, so choose a higher value;
- if lhs > rhs, then trial x/d is too high, so choose a lower value;
- iterate until satisfactory convergence is obtained for the equilibrium condition.

As discovered in earlier chapters, compression steel is usually required when the section size (eg, the depth) is restricted. Therefore, in the typical design problem for a doubly-reinforced rectangular section, the width b, effective depth d, and cover d' are known beforehand. What is generally required is the area of reinforcement (tension and compression) for a given ultimate moment of resistance. The design steps (when the rectangular-parabolic stress block is used) are:

- ensure that compression reinforcement is required by establishing that $x/d > 0.5$ if the beam is to be singly reinforced, or show that the equivalent ultimate resistance moment is less than that prescribed;
- assume a value for the compression steel ratio, say, between 0.5 and 4 per cent;
- determine the tension steel ratio ρ by combining equations (9.29) and

(9.35) with $\sigma_{st} = 0.87f_y$ and $\sigma_{sc} = 0.72f_y$.

It should be noted that the values of ρ and ρ' obtained from the foregoing will, most probably, not be optimal [such as to make the total reinforcement amount $(\rho + \rho')$ attain the minimum value]. It can be shown that the optimal values of ρ and ρ' will make the tension steel attain the limiting strain ϵ_2.

Therefore, an additional step may be to repeat the preceding steps by trying other values of ρ', computing ρ and then plotting a graph of $(\rho + \rho')$ versus ρ'. The optimal $(\rho + \rho')$ corresponds to the minimum point on the curve and the value of ρ' can easily be read off.

An alternative (and perhaps a more popular) approach to the above is to:

- guess a value for x;
- determine the strains ϵ_{st}, ϵ_{sc} in both tension and compression reinforcement using equations (9.31) and (9.32);
- obtain the stresses σ_{st}, σ_{sc} (in both tension and compression reinforcement) that correspond to the computed strains ϵ_{st} and ϵ_{sc}, using the design stress-strain curve (Fig. 9.3);
- compute ρ and ρ' by solving equations (9.30) and (9.35) simultaneously;
- to obtain optimal values of ρ and ρ', repeat the above procedure and plot a graph of $(\rho + \rho')$ versus x. Then read off the appropriate value of x and optimal $(\rho + \rho')$. Combine these with either equation (9.30) or equation (9.35) to obtain the required optimal ρ and ρ'.

If the design is to be based on the simplified stress block then the procedure is to:

- obtain the compression steel area from equation (9.36);
- determine the amount of tension steel from equation (9.37).

Example 9.3

A 300 mm × 600 mm beam is reinforced with three 32 mm ϕ steel bars in the tension region and two 25 mm ϕ bars in the compression zone. If the total cover for the bars is 40 mm in each region, determine the ultimate resistance moment of the section. Assume $f_{cu} = 30\ \text{N/mm}^2$ and $f_y = 410\ \text{N/mm}^2$.

Solution

$b = 300$ mm, $d = 600 - 40 = 560$ mm, $A_{st} = 2413\ \text{mm}^2$, $A_{sc} = 982\ \text{mm}^2$.
$\rho = 2413/(300 \times 560) = 1.44 \times 10^{-2}$;
$\rho' = 982/(300 \times 560) = 5.84 \times 10^{-3}$;
$d' = 40$ mm

As $f_{cu} = 30\ \text{N/mm}^2$, $K_1 = 0.403$ and $K_2 = 0.417$ try $x/d = 0.6$.
Then:

$$x = 336\ \text{mm}$$

From equations (9.31) and (9.32):

$$\epsilon_{st} = (1/0.6 - 1) \times 0.0035 = 0.0023$$

$$\epsilon_{sc} = (1 - 40/336) \times 0.0035 = 0.0031$$

But, from equations (9.13) and (9.14):

$$\epsilon_1 = 410/287\,500 = 0.0014$$

$$\epsilon_2 = 0.002 + 410/230\,000 = 0.0038$$

Therefore, $\epsilon_1 \leq \epsilon_{st} \leq \epsilon_2$

or, from equation (9.12):

$$\sigma_{st} = \left[4 + \frac{(0.0023 - 0.0014)}{(0.0038 - 0.0014)}\right](410/5.75) = 320 \text{ N/mm}^2$$

As $\epsilon_{sc} > 0.002$, $\sigma_{sc} = 0.72 \times 410 = 295 \text{ N/mm}^2$
The lhs of equation (9.30) $= 0.403 \times 30 \times 0.6 = 7.254$
The rhs of equation (9.30) $= 1.44 \times 10^{-2} \times 320 - 5.84 \times 10^{-3} \times 295$
$= 2.873$

Therefore; lhs > rhs and so the estimate for x/d was too high.
Try $x/d = 0.3$; $x = 168$ mm.

$$\epsilon_{st} = (1/0.3 - 1) \times 0.0035 = 0.0082$$

$$\epsilon_{sc} = (1 - 40/168) \times 0.0035 = 0.0027$$

Therefore, $\epsilon_{st} > \epsilon_2$ or $\sigma_{st} = 0.87 \times 410 = 357 \text{ N/mm}^2$
$\epsilon_{sc} > 0.002$ or $\sigma_{sc} = 295 \text{ N/mm}^2$
The lhs of equation (9.30) $= 0.403 \times 30 \times 0.3 = 3.627$
The rhs of equation (9.30) $= 1.44 \times 10^{-2} \times 357 - 5.84 \times 10^{-3} \times 295$
$= 3.418$

As the lhs is still a little higher than the rhs the estimate is still a little too high. Any other choice of $x/d < 0.3$ will still make $\sigma_{st} = 357 \text{ N/mm}^2$ and $\sigma_{sc} = 295 \text{ N/mm}^2$ because the critical strains have already been exceeded. With this in mind, simply find the value of x/d required by solving equation (9.30) directly (as the stresses are now known). Thus:

$$x/d = 3.418/(0.403 \times 30) = 0.28$$

or

$$x = 157 \text{ mm}$$

Using equation (9.35):

$$\frac{M_u}{bd^2} = 5.84 \times 10^{-3} \times 295[0.417 \times 0.28 - (40/560)] + 1.44 \times 10^{-2} \times 357[1 - 0.417 \times 0.28]$$

$$= 4.62 \text{ N/mm}^2$$

or

$$M_u = 434.5 \text{ kN m}$$

Example 9.4

A rectangular beam of width 280 mm and effective depth 580 mm is required to carry an ultimate moment of 470 kN m. Determine the amount of reinforcement required. Assume $f_{cu} = 30 \text{ N/mm}^2$, $f_y = 410 \text{ N/mm}^2$ and cover to reinforcement = 40 mm.

Solution

$K_1 = 0.403$, $K_2 = 0.417$, $b = 280$ mm, d = 580 mm.

If the beam were to be singly reinforced, the ultimate moment (which would correspond to $x/d = 0.5$, or $x = 290$ mm) would be given by equation (9.23) with $z = 0.75d$ or:

$$M_u = 0.403 \times 30 \times 290 \times 280 \times 0.75 \times 580$$
$$= 427 \text{ kN m}$$

which is less than the 470 kN m required for the beam in this example.

Therefore, compression reinforcement is required.

Assume $\rho' = 1$ per cent. From equation (9.30), with $\sigma_{st} = 0.87 \times 410 = 357\ \text{N/mm}^2$ and $\sigma_{sc} = 0.72 \times 410 = 295\ \text{N/mm}^2$:

$$x/d = (357\rho - 0.01 \times 295)/(0.403 \times 30)$$
$$= 29.5\rho - 2.44$$

From equation (9.35):

$$\frac{470 \times 10^6}{280 \times 580^2} = 5.0$$
$$= 0.01 \times 295[0.417(x/d) - (40/580)]$$
$$+ 357\rho[1 - 0.417(x/d)]$$

or:

$$1.23(x/d) + 357\rho[1 - 0.417(x/d)] = 5.02$$

Putting $(x/d) = 29.5\rho - 2.44$:

$$548\rho^2 - 94\rho + 1 = 0$$

from which $\rho = 1.1$ per cent or 16.0 per cent.

The second answer is unrealistic because it provides a value of $x/d > 1$. Therefore:

$$A_{st} = 0.011 \times 280 \times 580 = 1786\ \text{mm}^2$$
$$A_{sc} = 0.01 \times 280 \times 580 = 1624\ \text{mm}^2$$

If the simplified stress block is used, then from equation (9.36):

$$A_{sc} = \frac{470 \times 10^6 - 0.15 \times 30 \times 280 \times 580^2}{0.72 \times 410 \times (580-40)} = 289\ \text{mm}^2$$

while, from equation (9.37):

$$A_{st} = \frac{0.2 \times 30 \times 280 \times 580 + 0.72 \times 410 \times 289}{0.87 \times 410} = 2971\ \text{mm}^2$$

Comments

As noted earlier, the amount of reinforcement obtained on the basis of the rectangular-parabolic stress block is not necessarily the optimal value. The extra step required to obtain the optimal steel reinforcement is left as an exercise to the student.

9.1.3 Flanged sections

Flanged sections were introduced earlier in part 1 of this book. The two most common flanged sections are the T-beam and L-beam.

CP 110 recommends that, in the absence of a better estimate, the effective flange width for a T-beam should not exceed the lesser of:

- the width of the web plus one-fifth of the distance between the points of zero moment; or
- the actual width of the flange.

For an L-beam, CP 110 recommends that the effective flange width is limited to the lesser of:

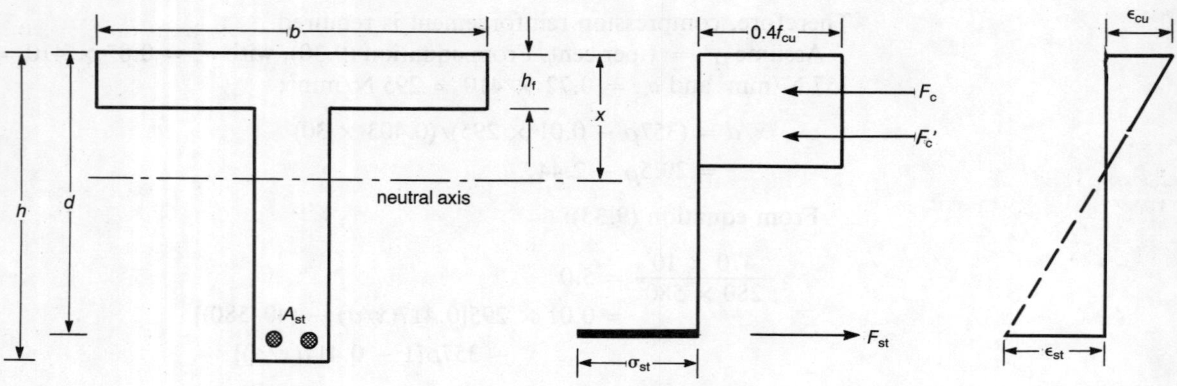

Fig. 9.5 T-beam section

- the width of the web plus one-tenth of the distance between the points of zero moment; or
- the actual width of the flange.

For a continuous beam, the Code allows the distance between the points of zero moment to be taken as 0.7 times the effective span.

As has been stated earlier in chapter 3, when the neutral axis falls within the flange the section behaves essentially as a rectangular beam, and all the formulae that have already been derived can be used. If the neutral axis is below the flange, a decision must be made as to whether or not to neglect the compression carried by the web.

It is obviously convenient to use the simplified stress block when designing a flanged beam. Now, consider the T-beam shown in Fig. 9.5. Let the neutral axis fall below the soffit of the flange, as shown, and let the simplified stress block be adopted. The compression carried by the flange is:

$$F_c = 0.4f_{cu}\, bh_f \tag{9.40}$$

while that in the rib is:

$$F_c' = 0.4f_{cu}\, b_w(x - h_f) \tag{9.41}$$

As the tension in the steel is $F_{st} = \sigma_{st} A_{st}$, the equilibrium condition $F_{st} = F_c + F_c'$ is given by:

$$\sigma_{st} A_{st} = 0.4f_{cu}b[h_f + b_w(x - h_f)/b] \tag{9.42}$$

in which σ_{st} depends on the neutral axis factor x/d.

The ultimate resistance moment can be obtained by taking moments about the location of reinforcement as:

$$\begin{aligned} M_u &= F_c(d - h_f/2) + F_c'[d - (x - h_f)/2] \\ &= 0.4f_{cu}bh_f(d - h_f/2) + 0.4f_{cu}b_w(x - h_f)[d - (x - h_f)/2] \end{aligned} \tag{9.43}$$

If the rib compression forces are neglected, the ultimate resistance moment based on concrete is:

$$M_u = 0.4f_{cu}\, bh_f(d - h_f/2) \tag{9.44}$$

and when based on steel (with $\sigma_{st} = 0.87f_y$):

$$M_u = 0.87f_y A_{st}(d - h_f/2) \tag{9.45}$$

Equations (9.44) and (9.45) are the simplified design formulae recommended in CP 110 and the ultimate resistance moment of a flanged beam is to be taken as the lesser of the two.

9.1.4 Non-rectangular sections

The derivation of the formulae for sections having non-rectangular geometries is along the line established above for rectangular and flanged sections. When the rectangular-parabolic stress curve is used, the analysis, although complex, is obtained by performing appropriate integrations across the section. The procedure is obviously less complicated when the simplified stress block is adopted. In such cases, the point of action of the total force in the compression zone is simply the centroid of the section itself. Exercises on the design/analysis of complicated sections are provided in section 9.5.

9.2 Limit state of deflection

One of the basic requirements of the serviceability limit state is that the deflection of the structure (or any part of it) should not affect adversely the appearance or efficiency of the structure (CP 110, 1972). The Code only provides a guide to what are considered 'reasonable limits' of deflection. The engineer is saddled with the responsibility of determining satisfactory conditions for which the deflections are not excessive.

As a guide to limiting the deflection the Code requires that:

- the final deflection (including the effects of temperature, creep and shrinkage), measured below the as-cast level of the supports of floors, roofs and all other horizontal members, should not, in general, exceed span/250;
- partitions and finishes will be affected only by that part of the deflection (including the effects of temperature, creep and shrinkage) that takes place after the construction of the partitions or the application of the finishes. Information is lacking, but a limit of span/350 or 20 mm (whichever is the lesser) is suggested.

Thus, if ξ_1 is the deflection before the construction of partitions or the application of finishes, and ξ_2 is the deflection after, then:

$$\xi_1 + \xi_2 \leq \text{span}/250 \tag{9.46}$$

and

$$\xi_2 \leq \text{span}/350 \text{ or } 20\,\text{mm (whichever is the lesser)} \tag{9.47}$$

In complying with the above requirements, the designer can either:

- calculate the deflection of the beam and ensure that the conditions given by equations (9.46) and (9.47) are not violated (suggestions as to how these calculations can be made are given in Appendix A.2 of the Code);

or in all normal cases (for rectangular and flanged beams):

- simply ensure that the span/depth ratios do not exceed certain values which are specified in the Code and which are quoted below.

(i) Rectangular beams. The basic span/effective depth ratios are shown in Table 9.1. The ratios are based on compliance with equation (9.46) and, for beams of up to 10 m span, damage to finishes and partitions should normally be avoided under these conditions.

Table 9.1 Basic span/effective depth ratios for rectangular beams

Support conditions	Ratio
cantilever	7
simply supported	20
continuous	26

Table 9.2 Special span/effective depth ratios for rectangular beams

Support conditions	Span (m)				
	12	14	16	18	20
cantilever	value to be justified by calculation				
simply supported	18	16	14	12	10
continuous	23	21	18	16	13

Table 9.3 Modification factors for tension reinforcement

Service stress (σ_{st}) (N/mm²)	100 A_{st}/bd							
	0.25	0.50	0.75	1.00	1.50	2.00	2.50	≥3.0
145 (f_y = 250)	2.0	1.98	1.62	1.44	1.24	1.13	1.06	1.01
150	2.0	1.91	1.58	1.41	1.22	1.11	1.04	0.99
200	2.0	1.46	1.26	1.15	1.02	0.94	0.89	0.85
238 (f_y = 410)	1.60	1.23	1.09	1.00	0.90	0.84	0.80	0.77
246 (f_y = 425)	1.55	1.20	1.06	0.98	0.88	0.83	0.79	0.76
250	1.52	1.18	1.05	0.97	0.87	0.82	0.78	0.75
267 (f_y = 460)	1.41	1.11	0.99	0.92	0.84	0.78	0.75	0.72
290 (f_y = 500)	1.27	1.03	0.92	0.86	0.79	0.74	0.71	0.68
300	1.22	0.99	0.90	0.84	0.77	0.72	0.69	0.67

The Code recommends the ratios given in Table 9.2 for spans exceeding 10 m when it is necessary to restrict the deflection below the span/250 limit (on which Table 9.1 is based).

As deflection does not depend solely on the section size but also on the amount of tension reinforcement and its stress, the span/effective depth ratios must be modified according to the area of reinforcement provided and its service stress at the centre of the span (or at the support in the case of a cantilever).

The modification factors that should be used to multiply the span/effective depth ratios in Tables 9.1 and 9.2 are shown in Table 9.3.

The service stress σ_{st} can be obtained from:

$$\sigma_{st} = 0.58 f_y A_0 / \beta_b A_1 \tag{9.48}$$

in which f_y is the characteristic strength of the reinforcement; A_0 is the area of reinforcement required at mid-span (or at the support for a cantilever) to resist the moment due to ultimate loads; A_1 is the area of reinforcement provided at mid-span (or at the support for a cantilever); and β_b is the ratio of the resistance moment (provided at mid-span) to the mid-span moment (taken from the maximum-moments diagram before redistribution).

Table 9.4 Modification factors for compression reinforcement

$100\ A_{sc}/bd$	Factor
0.25	1.07
0.50	1.14
0.75	1.20
1.0	1.25
1.5	1.33
2.0	1.40
≥3.0	1.50

In the event that the percentage of redistribution is not known but the design ultimate moment at mid-span is at least equal to the elastic ultimate moment, then the stress σ_{st} in Table 9.3 may be taken as $0.58f_y$.

The appropriate modification factors to account for the influence of compression reinforcement on deflection are shown in Table 9.4. Intermediate values can be obtained by simple linear interpolation. The area A_{sc} of compression reinforcement used in Table 9.4 may include all bars in the compression zone, even those not tied with links.

Normal creep and shrinkage deflection have been adequately accounted for in the permissible span/effective depth ratios obtained from Tables 9.1 to 9.4. However, if there is reason to expect that creep or shrinkage of the concrete may be high (such as when the free shrinkage strain >0.075 per cent or the creep coefficient >4), or if some other adverse conditions are expected, the allowable span/effective depth ratios should be reduced accordingly.

(ii) Flanged beams. The Code recommends the use of Tables 9.1 to 9.4 for obtaining the span/effective depth ratio. However, if the web width is less than 0.3 times the effective flange width, the final ratio obtained should be multiplied by 0.8.

When the value of the web width (b_w) to the effective flange width (b) is greater than 0.3, the multiplication factor is given by:

$$\alpha = 2/7(b_w/b) + 5/7 \tag{9.49}$$

Reinforcement percentages in flanged sections are to be based on the effective flange width b of the beam, multiplied by the effective depth d.

For beams having more complicated cross-sections (or whenever the designer wishes to adopt this approach), the deflection can be calculated by solving the differential equation:

$$\frac{d^2y}{dx^2} = 1/r_x \tag{9.50}$$

in which $1/r_x$ is the curvature at a distance x from the origin. When closed-form solutions are not possible, a method based on numerical integration must be used.

In general, the deflection a is given by

$$a = Kl^2/r_b \tag{9.51}$$

in which l is the effective span of the member, $1/r_b$ is the curvature at mid-span (or, for cantilevers, at the support section), and K is a constant that is a function of the shape of the bending moment diagram.

As outlined in Appendix A.2 of the Code, the curvature may be calculated by assuming that either:

(A) the section is cracked under the loading considered; or

(B) the section is uncracked.

The assumption yielding a larger value of the curvature is the one that should be adopted.

Under the assumption (A) [quoting section A.2.2 of the Code]:

- strains are calculated on the assumption that plane sections remain plane;
- the reinforcement, whether in tension or in compression, is assumed elastic and its modulus of elasticity may be taken as 200 kN/mm^2;
- the concrete in compression is assumed to be elastic. Under short-term loading, the modulus of elasticity may be taken as that provided in Table 8.1. Under long-term loading, an effective modulus may be taken as having a value of $1/(1+\phi)$ times the short term modulus, where ϕ is the appropriate creep coefficient;
- stresses in the concrete in tension may be calculated on the assumption that the stress distribution is triangular, having a value of zero at the neutral axis and a value at the centroid of the tension steel of 1 N/mm^2 instantaneously, reducing to 0.55 N/mm^2 in the long term [see Exercise 9.9].

Under the assumption (B), all the points listed above for (A) (with the exception of part of the fourth) are applicable. The stress in concrete at the centroid of the tension steel should not be greater than 1.0 N/mm^2.

In order to assess the total long-term curvature of a section, the Code recommends the following procedure:

- calculate the instantaneous curvatures under the total load and under the permanent load;
- calculate the long-term curvature under the permanent load;
- add to the long-term curvature under the permanent load the difference between the instantaneous curvatures under the total and permanent loads;
- add to this curvature the shrinkage curvature calculated from the formula:

$$1/r_{cs} = \rho_0 \, \epsilon_{cs}/d \tag{9.52}$$

in which $1/r_{cs}$ is the shrinkage curvature; ϵ_{cs} is the free shrinkage strain (in the assessment of which allowance should be made for the shape of the section); d is the effective depth of the member; and ρ_0 is a coefficient that depends on the percentages of tension and compression steel in the section. Values of ρ_0 given in the Code are quoted in Table 9.5.

Example 9.5 The doubly-reinforced section of Example 9.3 is that of a simply-supported beam of effective span 9.5 m. Check that the serviceability limit state of deflection is not violated, given that the A_{st} required is 2401 mm^2.

Solution From Example 9.3; $f_{cu} = 30$ N/mm^2, $f_y = 410$ N/mm^2, $d = 560$ mm, $b = 300$ mm, $A_{sc} = 982$ mm^2; A_{st} (provided) $= 2413$ mm^2.

Therefore, $100\,A_{sc}/bd = 0.58$, $100\,A_{st}/bd = 1.44$

Using equation (9.48) with $\beta_b = 1$:

$$\sigma_{st} = 0.58 \times 410 \times 2401/2413 = 237 \text{ N/mm}^2$$

$100\,A_{st}/bd$	$100\,A_{sc}/bd$								
	0.00	0.25	0.50	0.75	1.00	1.25	1.50	1.75	2.00
0.25	0.44	0.31	0.26	0.22	0.20	0.18	0.17	0.16	0.15
0.50	0.56	0.31	0.26	0.22	0.20	0.18	0.17	0.16	0.15
0.75	0.64	0.45	0.26	0.22	0.20	0.18	0.17	0.16	0.15
1.00	0.70	0.55	0.39	0.22	0.20	0.18	0.17	0.16	0.15
1.50	0.80	0.69	0.57	0.45	0.32	0.18	0.17	0.16	0.15
2.00	0.88	0.79	0.69	0.60	0.49	0.39	0.28	0.16	0.15
2.50	0.95	0.87	0.79	0.70	0.62	0.53	0.44	0.35	0.25
3.00	1.00	0.94	0.86	0.79	0.72	0.64	0.57	0.49	0.40
3.50	1.00	1.00	0.93	0.87	0.80	0.74	0.67	0.60	0.52
4.00	1.00	1.00	1.00	0.93	0.87	0.81	0.75	0.69	0.62

Table 9.5 Values of ρ_0 for calculation of shrinkage curvatures

From Table 9.1, maximum span/effective depth allowed = 20.
From Table 9.3, the modification factor for tension reinforcement = 1.264 (by interpolation).
Additional modification factor for compression reinforcement (from Table 9.4) = 1.16.
Hence the permissible span/effective depth = $1.264 \times 1.16 \times 20 = 29.3$.
But, the actual span/effective depth = $9.5 \times 10^3/560 = 17.0 < 29.3$.
Therefore, the serviceability limit state of deflection is satisfied.

9.3 Limit state of cracking

The basic requirement of the serviceability limit state of cracking is to ensure that the appearance or efficiency of the structure will not be affected adversely by the cracking of the concrete. The designer must be satisfied that the structure (having due regard for the particular environment in which it is located) will not suffer excessive cracking. As has been stated (more or less) earlier, the Code provides a guide to what are considered 'reasonable' limits of crack widths. These acceptable limits are:

- the surface crack widths, when assessed, should not in general exceed 0.3 mm;
- in cases where the environment can be considered aggressive (eg, sea water) the assessed surface crack widths at points nearest the main reinforcement should not exceed 0.004 times the nominal cover to the main reinforcement (see Table 9.6).

The development of cracks in a reinforced concrete structure cannot be adequately studied using analytical or deterministic techniques only, because there exists a random component in the process. The engineer must therefore settle for an acceptable probability that the predicted maximum crack width will not be exceeded.

In complying with the requirements of the serviceability limit state of cracking, the designer can either:

- compute the crack widths and ensure that the largest of these do not exceed the recommended allowable values quoted above; or
- simply ensure that the rules for reinforcement spacing, provided in the Code, are followed.

Table 9.6 Nominal cover to reinforcement (mm)

Condition of exposure	Concrete grade				
	20	25	30	40	50 and over
Mild: eg, completely protected against weather or aggressive conditions, except for brief period of exposure to normal weather conditions during construction.	25	20	15	15	15
Moderate: eg, sheltered from severe rain and against freezing whilst saturated with water. Buried concrete and concrete continuously under water.	—	40	30	25	20
Severe: eg, exposed to driving rain, alternate wetting and drying and to freezing whilst wet. Subject to heavy condensation or corrosive fumes.	—	50	40	30	25
Very severe: eg, exposed to sea water or moorland water and with abrasion.	—	—	—	60	50
Subject to salt used for de-icing:	—	—	50*	40*	25

*only applicable if the concrete has entrained air

(i) Calculation of crack widths: There are three major factors influencing the widths of cracks at a given point for members undergoing bending. These are:

- the proximity to the point of reinforcing bars perpendicular to the cracks;
- the proximity of the neutral axis to the point;
- the average surface strain at the point considered.

Under the proviso that the strain in the tension reinforcement is limited to $0.8f_y/E_{st}$, the formula (given in the Code) for the critical crack width ξ, which takes the above three factors into account, is:

$$\xi = \frac{3\,a_{cr}\,\epsilon_m}{1 + 2(a_{cr} - c_{min})/(h - x)} \tag{9.53}$$

in which a_{cr} is the distance from the point considered to the surface of the nearest longitudinal bars; ϵ_m is the average strain at the level where cracking is considered; c_{min} is the minimum cover to the tension steel; and x is the depth of the neutral axis. The strain ϵ_m is to be determined from the expression:

$$\epsilon_m = \epsilon_1 - \frac{1.2\,b_t\,h(a' - x)}{1000A_{st}(h - x)f_y} \tag{9.54}$$

f_y(N/mm²)	percentage redistribution to or from section considered										
	−30	−25	−20	−15	−10	0	+10	+15	+20	+25	+30
200	215	230	245	260	275	300	300	300	300	300	300
410	130	140	150	155	165	185	205	215	220	230	240
425	125	135	145	155	160	180	200	210	215	225	235
460	115	125	130	140	150	165	180	190	200	205	215
500	105	115	120	130	135	150	165	175	180	190	195

Table 9.7 Clear distance between bars (mm)

where ϵ_1 is the strain at the level considered, calculated by ignoring the stiffening effect of the concrete in the tension zone; b_t is the width of the section at the centroid of the tension steel; a' is the distance from the compression face to the point at which the crack width is being calculated; and A_{st} is, of course, the area of tension reinforcement. A section is uncracked if ϵ_m has a negative value.

(ii) Reinforcement spacing rules: The following rules (quoted from CP 110) are to be applied to beams in normal internal or external conditions of exposure (unless the calculation of crack widths reveals that a larger spacing is acceptable):

- in the application of these rules any bar with a diameter less than 0.45 times the diameter of the maximum bar size in the section should be ignored, except when considering those in the side faces of beams. Bars placed in the side faces of beams to control cracking should be of a size not less than $\sqrt{(s_b b/f_y)}$, where s_b is the spacing of the bars and b the breadth of the section at the point considered;
- the clear horizontal distance between adjacent bars, or groups, near the tension face of a beam should not be greater than the value given in Table 9.7, depending on the amount of redistribution carried out in analysis and the characteristic strength of the reinforcement;
- the clear distance from the corner of a beam to the surface of the nearest longitudinal bars should not be greater than half the clear distance given in Table 9.7;
- when the overall depth of a beam exceeds 750 mm, longitudinal bars should be provided over a distance of 2/3 of the overall depth from the tension face. This reinforcement should be positioned near the side faces and be spaced at not more than 250 mm; it may be used in calculating the resistance moment of the section.

The rules cannot be applied to structural members in an aggressive environment, unless the characteristic yield stress f_y in steel is limited to 300 N/mm² in the calculation of the resistance moment.

9.4 Shear

A general account of shear forces in beams has been given earlier in chapter 3. In the limit state method the shear stress v at any cross-section in a beam can be obtained using the expression:

$$v = V/bd \tag{9.55}$$

in which V is the shear force due to ultimate loads; b is the breadth of the

Table 9.8 Ultimate shear stress in beams (N/mm²)

$100A_{st}/bd$	Concrete grade			
	20	25	30	≥40
0.25	0.35	0.35	0.35	0.35
0.50	0.45	0.50	0.55	0.55
1.00	0.60	0.65	0.70	0.75
2.00	0.80	0.85	0.90	0.95
3.00	0.85	0.90	0.95	1.00

Table 9.9 Maximum values of shear stress in beams

Concrete grade	Maximum shear stress (N/mm²)
20	3.35
25	3.75
30	4.10
≥40	4.75

section which, for a flanged beam, should be taken as the rib width, and d is the effective depth.

In the event that the shear stress exceeds the ultimate shear stress (Table 9.8) given in the Code, shear reinforcement must be provided. The shear stress v cannot exceed the values given in Table 9.9, even when shear reinforcement is provided.

The spacing of the legs of links (when used for shear reinforcement) in the direction of the span and perpendicular to it, has been limited to $0.75d$ in CP 110. Furthermore, the spacing s_v of links along the member must be such that:

$$s_v \leq \frac{0.87 f_{yv} A_{sv}}{b(v - v_c)} \tag{9.56}$$

in which f_{yv} is the characteristic strength of the link reinforcement (which should not be taken as more than 425 N/mm²); A_{sv} is the cross-sectional area of the two legs of a link; and v_c is the ultimate shear stress in concrete.

9.5 Torsion

As earlier stated in section 3.4, CP 110 provides guidelines (clause 3.3.7) to be followed in the design of beams for torsion. When the torsional resistance of a beam has not been taken into account in the analysis of the structure, the nominal reinforcement required for shear is considered by the Code to be adequate for the control of torsional cracking.

Where it is considered that the applied torsional moment is significant, the Code requires, for rectangular beams, that the torsion shear stress v_t be calculated from the equation:

$$v_t = \frac{2T}{h_{min}^2(h_{max} - \tfrac{1}{3}h_{min})} \tag{9.57}$$

where T is the torsional moment due to ultimate loads and h_{min} and h_{max} are, respectively, the smaller and larger dimensions of the section. For flanged beams (T-, L- or I-sections) the calculation is carried out by subdividing the sections into their component rectangles.

The subdivision should be such that it maximises the sum of $h_{min}^3 h_{max}$ over all component rectangles. As suggested in the Code, the maximisation is generally achieved when the widest component rectangle is made as long as possible. The torsion shear stress is still given by equation (9.57) but the torsional moment carried by the jth component rectangle is now:

$$T_j = T \frac{(h_{min}{}^3 h_{max})_j}{\sum_{j=1}^{N} (h_{min} \; h_{max})_j} \tag{9.58}$$

in which N is the total number of component rectangles.

The recommended values of ultimate torsion shear stress given by the Code are quoted in Table 9.10.

Table 9.10 Ultimate torsion shear stress

	Concrete grade			
	20	25	30	≥40
v_{tmin} (N/mm²)	0.30	0.33	0.37	0.42
v_{tu} (N/mm²)	3.35	3.75	4.10	4.75

Whenever $v_t > v_{min}$ torsion reinforcement should be provided. The combined shear stresses due to shear force and torsion should not exceed v_{tu}.

Reinforcement for torsion is to consist of rectangular closed links, together with longitudinal reinforcement. The torsional reinforcement A_{sv} is to be considered additional to any reinforcement required for shear or bending. The criteria recommended for the calculation of A_{sv} are:

$$\frac{A_{sv}}{s_v} \geq \frac{T}{0.8 x_1 y_1 (0.87 f_{yv})} \tag{9.59}$$

$$A_{sl} \geq \frac{A_{sv}}{s_v} \left(\frac{f_{yv}}{f_{yl}} \right) (x_1 + y_1) \tag{9.60}$$

where A_{sv} is the area of the legs of closed links at a section; A_{sl} is the area of longitudinal reinforcement; f_{yv} is the characteristic strength of the links; f_{yl} is the characteristic strength of the longitudinal reinforcement; s_v is the spacing of the links; x_1 is the smaller dimension of the links, and y_1 is the larger dimension of the links.

The maximum allowable value of either f_{yv} or f_{yl} that can be used in the above criteria is 425 N/mm². The spacing of the links is not to exceed the least of:

- x_1;
- $y_1/2$;
- 200 mm.

In sections where $y_1 < 550$ mm (ie, small sections) the torsion shear stress v_t is not to exceed $v_{tu} \, y_1/550$.

9.6 Exercises

9.1 A 300 mm × 700 mm rectangular beam is reinforced with four 32 mm ϕ steel bars. The cover provided is 50 mm. Determine the ultimate resistance moment of the section if the characteristic strength in concrete f_{cu} is 40 N/mm² and that in steel f_y is 410 N/mm².

9.2 Using the rectangular-parabolic stress block, determine the depth d and reinforcement area A_{st} required for a 420 mm wide rectangular beam, if the ultimate moment of resistance is 308 kN m. Assume f_{cu} = 30 N/mm² and f_y = 300 N/mm².

9.3 Attempt Exercise 9.2 again, using the simplified stress block.

9.4 Determine the appropriate section size and reinforcement for a square beam required to carry an ultimate resistance moment of 308 kN m. Assume f_{cu} = 30 N/mm² and f_y = 300 N/mm².

9.5 A 300 mm × 800 mm beam is reinforced with four 32 mm ϕ steel bars in the tension zone and two 20 mm ϕ bars in the compression region. If the cover to the reinforcement is 50 mm, what is the ultimate moment of resistance of the section? Assume f_{cu} = 30 N/mm² and f_y = 450 N/mm².

9.6 A 300 mm wide rectangular beam with an overall depth of 600 mm is to carry an ultimate resistance moment of 520 kN m. Determine the amount of reinforcement required. Assume f_{cu} = 40 N/mm², f_y = 410 N/mm² and that the cover to the reinforcement = 40 mm.

9.7 A T-beam is required to carry an ultimate moment of 430 kN m. If the flange width = 950 mm, the flange thickness = 150 mm, and the rib width = 280 mm, determine the appropriate overall depth and reinforcement.

9.8 Attempt Exercise 3.13 using the limit state method. From the results produce appropriate charts that could be used in the design office for the design/analysis of these sections.

9.9 Obtain the expression for the resistance moment of a cracked singly-reinforced rectangular beam, assuming that the stress distribution is triangular and the maximum tensile stress allowed in the concrete is σ_{ct}.

9.10 Check that the serviceability limit state of deflection is satisfied for the beam in Exercise 9.6, given that the section is that of a simply-supported member of effective span 12.4 m.

9.11 Calculate the maximum surface crack widths in the beams of Exercise 9.1 and 9.5.

9.12 Using the bar spacing rules, check whether the serviceability limit state of cracking is violated by the beam sections of Exercises 9.1 and 9.5.

10 Slabs and bases

10.1 Introduction

A detailed description of the common types of slabs and bases has been presented earlier in chapters 1, 4 and 7. The discussion in this chapter will therefore be limited to the special considerations required in the design of slabs/bases when the limit state method of design is used. In general, a significant amount of the previous treatment of beams (in chapter 9) will also be applicable to slabs and bases.

Slabs spanning in one direction are designed in just the same way as beams by considering a unit strip of width 1000 mm. The ultimate resistance moment of a cross-section in a solid slab is determined using the methods established in section 9.1.

10.2 Solid slabs spanning in two directions

Consider a solid slab spanning in two directions (orthogonal to each other) and carrying a uniformly-distributed ultimate load intensity of amount w. The two special cases will be when: (a) the slab is simply supported on all edges (with no provision for torsion and lifting); and (b) the slab is somehow restrained, with provision to resist torsion at the corners and to prevent the corners from lifting.

10.2.1 Simply-supported solid slabs

The Code (clause 3.4.3.1) recommends that the maximum moments per unit width be obtained as:

$$M_x = \alpha_x w l_x^2 \tag{10.1}$$

$$M_y = \alpha_y w l_x^2 \tag{10.2}$$

in which M_x and M_y are the moments at mid-span on strips of unit width and spans l_x and l_y, respectively; w = (1.4 × dead load intensity + 1.6 × live load intensity); l_y and l_x are the lengths of the longer and shorter sides respectively; and α_x and α_y are the same moment coefficients shown earlier in Table 4.1.

Not less than half of the tension reinforcement provided at mid-span should extend to the supports. The remaining half must extend to within $l_x/10$ or $l_y/10$ of the support, as appropriate.

Table 10.1 Bending moment coefficients for rectangular panels supported on four sides with provision for torsion at the corners

Case	Types of panel and moments considered	Short span coefficients β_x, l_y / l_x = 1.0	1.1	1.2	1.3	1.4	1.5	1.75	2.0	Long span coefficients β_y for all values of l_y / l_x
1	**Interior panels:**									
	Negative moment at continuous edge	0.032	0.037	0.043	0.047	0.051	0.053	0.060	0.065	0.032
	Positive moment at mid-span	0.024	0.028	0.032	0.036	0.039	0.041	0.045	0.049	0.024
2	**One short edge discontinuous:**									
	Negative moment at continuous edge	0.037	0.043	0.048	0.051	0.055	0.057	0.064	0.068	0.037
	Positive moment at mid-span	0.028	0.032	0.036	0.039	0.041	0.044	0.048	0.052	0.028
3	**One long edge discontinuous:**									
	Negative moment at continuous edge	0.037	0.044	0.052	0.057	0.063	0.067	0.077	0.085	0.037
	Positive moment at mid-span	0.028	0.033	0.039	0.044	0.047	0.051	0.059	0.065	0.028
4	**Two adjacent edges discontinuous:**									
	Negative moment at continuous edge	0.047	0.053	0.060	0.065	0.071	0.075	0.084	0.091	0.042
	Positive moment at mid-span	0.035	0.040	0.045	0.049	0.053	0.056	0.063	0.069	0.035
5	**Two short edges discontinuous:**									
	Negative moment at continuous edge	0.045	0.049	0.052	0.056	0.059	0.060	0.065	0.069	—
	Positive moment at mid-span	0.035	0.037	0.040	0.043	0.044	0.045	0.049	0.052	0.035
6	**Two long edges discontinuous:**									
	Negative moment at continuous edge	—	—	—	—	—	—	—	—	0.045
	Positive moment at mid-span	0.035	0.043	0.051	0.057	0.063	0.068	0.080	0.088	0.035
7	**Three edges discontinuous (one long edge continuous):**									
	Negative moment at continuous edge	0.057	0.064	0.071	0.076	0.080	0.084	0.091	0.097	—
	Positive moment at mid-span	0.043	0.048	0.053	0.057	0.060	0.064	0.069	0.073	0.043
8	**Three edges discontinuous (one short edge continuous):**									
	Negative moment at continuous edge	—	—	—	—	—	—	—	—	0.057
	Positive moment at mid-span	0.043	0.051	0.059	0.065	0.071	0.076	0.087	0.096	0.043
9	**Four edges discontinuous:**									

10.2.2 Restrained solid slabs

For slabs that are adequately restrained at the corners, the recommended maximum moments per unit width are:

$$M_x = \beta_x w l_x^2 \qquad (10.3)$$

$$M_y = \beta_y w l_x^2 \qquad (10.4)$$

where β_x and β_y are the moment coefficients given in Table 10.1 (*cf.* Table 4.2).

The other highlights of the requirements of CP 110 in respect of restrained slabs are:

- where $l_y/l_x > 2$ the slab should be spanned in one direction only;
- the slabs are subdivided into middle and edge strips, as in Fig. 4.1;
- the maximum moments calculated from equations (10.3) and (10.4) apply only to the middle strips, and the moments should not be redistributed;
- the tension reinforcement provided at mid-span must extend, in the lower part of the slab, to within $l/4$ of a continuous edge, or $3l/20$ of a discontinuous edge. Not less than half of this reinforcement should extend to within $3l/20$ of a continuous edge, and to within 50 mm of a discontinuous edge.
- over the continuous edge of a middle strip, the tension bars should extend, in the upper part of the slab, a distance $3l/20$ from the support, and not less than half should extend a distance of $3l/10$.
- to take care of negative moments that may develop at a discontinuous edge, it is sufficient to provide tension reinforcement, the amount of which is half that provided at mid-span, extending $l/10$ into the span.
- at any corner where the slab is simply supported (on both edges meeting at the corner), torsion reinforcement (consisting of top and bottom bars) should be provided. The torsion reinforcement should be placed in layers parallel to the sides of the slab and extend at least $l_x/5$ from the edges. The area of reinforcement in each of the layers should be $\frac{3}{4} A_{sb}$, where A_{sb} is the required area of reinforcement for the maximum mid-span moment;
- at a corner contained by edges over only one of which the slab is continuous, torsion reinforcement equal to $\frac{3}{8} A_{st}$ should be provided;
- no torsion reinforcement is required at any corner contained by edges over both of which the slab is continuous;
- the loads on beams supporting solid slabs spanning in two directions may be assumed to be as shown in Fig. 4.2.

10.3 Flat slabs

The analysis of flat slabs using CP 110 follows closely the procedure established earlier in chapter 4 when the requirements of NCP 1 were being described. The requirements of CP 110 are that the following arrangement of load be considered:

- alternate spans loaded with total ultimate load ($1.4G_k + 1.6Q_k$), and all other spans loaded with minimum dead load ($1.0G_k$);
- any two adjacent spans loaded with total ultimate load ($1.4G_k + 1.6Q_k$), and all other spans loaded with minimum dead load ($1.0G_k$).

The division of the flat slab panels into column and middle strips is exactly as described to comply with NCP 1 (see Fig. 4.4). Allowance is made in CP 110 for analysis based on either the continuous-frame or the empirical method.

Shear walls or bracing must be provided for the purpose of resisting all lateral forces. When the empirical method is used the design moments for the strips and the supports are expressed as percentages of the moment M_0 which is given by:

$$M_0 = \frac{wL_2}{8}\left(L_1 - \frac{2D}{3}\right)^2 \qquad (10.5)$$

where L_1 is the length of a panel in the direction of span, measured from the centres of the columns; L_2 is the width of a panel measured from the centres of the columns; D is the diameter of the column or column head (see Fig. 4.5), which must be taken as the diameter of a circle of the same area as the cross-section of the head; and $w = (1.4g_k + 1.6q_k)$ is the total ultimate load per unit area of the panel.

The apportionment of moments between the column and middle strips and the supports is given in Table 10.2 which, with the exception of the wall supports, is identical to Table 4.7.

Table 10.2 Distribution of moments in flat slabs designed by the empirical method

	Apportionment of moments between the column and middle strips, expressed as percentages of M_0 (see **Note 3**)			
	Column strip (see **Note 2**)		Middle strip	
Interior panels				
With drops				
Negative moments	50		15	
Positive moments	20		15	
Without drops				
Negative moments	46		16	
Positive moments	22		16	
	Column supports	Wall supports	Column supports	Wall supports
Exterior panels (see **Note 1**)				
With drops				
Exterior negative moments	45	6	10	6
Positive moments	25	36	19	26
Interior negative moments	50	75	15	22
Without drops				
Exterior negative moments	41	6	10	6
Positive moments	28	40	20	28
Interior negative moments	46	66	16	24

Note 1. The percentages given in the first column are appropriate to conditions where the edge of the slab is supported on columns. Where the edge of the slab is supported by masonry, or other walls providing negligible restraint, the percentages given in the second column may be used.
Note 2. Where the column strip is taken as equal to the width of the drop, and the middle strip is thereby increased in width to a value greater than half the width of the panel, the moments to be resisted by the middle strip should be increased in proportion to its increased width. The moments to be resisted by the column strip may then be decreased by an amount such that there is no reduction in either the total positive or the total negative moments resisted by the column strip and middle strip together.
Note 3. Where adjacent spans differ in length, the longer span should be used to determine the negative moments.

The minimum thickness recommended by CP 110 for a flat slab is 125 mm. The lengths of adjoining panels should not differ by more than 15 per cent.

10.4 Ribbed, hollow block or voided slabs

The requirements of CP 110 are, with some minor differences, virtually identical to those of NCP 1 (see section 4) for ribbed, hollow block or voided slab constructions. The crushing strength of hollow blocks and forms must be at least 14 N/mm^2 (*cf.* 17.5 N/mm^2 for NCP 1) measured on the nett section, when axially loaded in the direction of compressive stress in the slab.

The thickness of topping, when used, should not be less than:

- 30 mm for slabs with permanent blocks and a clear distance between ribs of not more than 500 mm;
- 25 mm for slabs as just described but with each row of blocks jointed using cement-sand mortar not weaker than 1:3 or 11 N/mm^2;
- 40 mm or one-tenth of the clear distance between ribs, whichever is the greater, for all other slabs containing permanent blocks;
- 50mm or one-tenth of the clear distance between ribs, whichever is the greater, for all other slabs without permanent blocks.

10.5 Limit state of deflection

The guidelines established earlier in section 9.2 for beams are also applicable to solid slabs. Thus, the deflections may be computed and checked with the requirements of the serviceability limit state of deflection [see equations (9.46) and (9.47)]. Otherwise, the engineer may simply ensure that the span/effective depth is restricted to the allowable values. For solid slabs, the pertinent ratios and modification factors can be read from Tables 9.1 to 9.4.

CP 110 stipulates that only the reinforcement at the centre-span in the width of the solid slabs under consideration should be taken as influencing the deflection. The span and reinforcement along the shorter side will be the basis for the selection of the appropriate ratios for solid slabs that span in two directions.

For flat slabs having drops whose total width in both directions is at least equal to one-third of the respective spans, the provisions for solid slabs can be used. Otherwise, the basic span/effective depth ratios (required for equivalent solid slabs) should be multiplied by 0.9.

10.6 Limit state of cracking

The rules in CP 110 governing reinforcement spacing (which were set out earlier in section 9.3) will generally be adequate for the purpose of controlling flexural cracking in slabs. It may be advantageous, in some cases, actually to calculate the crack widths (using the formulae established in chapter 9) and then ensure that the values do not exceed the allowable amounts given in section 9.3.

10.7 Shear in slabs

As for beams, the shear stress v at a given cross-section in a solid slab can be calculated from the expression:

$$v = V/bd \tag{10.6}$$

in which V is the shear force due to ultimate loads; b is the width of the slab; and d is the effective depth. If the shear stress v is less than $\xi_s v_c$ (where ξ_s is given in Table 10.3 and v_c is the ultimate shear stress shown in Table 9.8), then no shear reinforcement will be necessary.

Table 10.3 Values of factor ξ_s

Overall slab depth h(mm)	Factor ξ_s
≥ 250	1.00
225	1.05
200	1.10
175	1.15
≤ 150	1.20

The requirements of CP 110 are that solid slabs with an overall depth $h < 200$ mm should not have shear stresses which exceed $\xi_s v_c$. Shear reinforcement must be provided for solid slabs with $h \geq 200$ mm when $v > \xi_s v_c$. In such cases the reinforcement will be as provided for beams, except that the spacing for links can be increased to the effective depth d (from $0.75d$ for beams).

For solid slabs carrying concentrated loads, the critical section for calculating shear is to be taken on a perimeter $1.5h$ from the boundary of the loaded area. In such slabs, when $h \geq 200$ mm and $\xi_s v_c \leq v \leq 1/2$ (the maximum allowable, from Table 9.9, for beams) then shear reinforcement must be provided on the critical perimeter, and a similar amount must be provided on the parallel perimeter, at a distance of $0.75h$ inside it, such that:

$$0.4\,\text{N/mm}^2 \leq \frac{\Sigma A_{sv}(0.87 f_y v)}{U_{crit}\, d} \geq v - v_c \tag{10.7}$$

in which ΣA_{sv} is the total area of shear reinforcement; $f_y v$ is the characteristic strength of the shear reinforcement, which should not be taken to be greater than 425 N/mm²; and U_{crit} is the length of the critical perimeter. Shear stress computations should be made on perimeters $0.75h$ out from the critical perimeter, and adequately reinforced until the value $\xi_s v_c$ is no longer exceeded.

For flat slabs (when the empirical method is used) a procedure identical to the preceding one (for solid slabs under concentrated loads) can be adopted by taking the column head as a concentrated load acting upwards. However, the values of the ultimate shear stress given in Table 9.8 must be reduced by 20 per cent.

In ribbed, hollow block or voided slabs, the shear calculation follows the pattern established for beams [equation (9.55)] with $b = b_w$, the width of the rib.

10.8 Staircases

The design of staircases must conform with the appropriate requirements established for beams and slabs. The ultimate load is generally taken as being distributed uniformly over the plan area of the staircase. For staircases with open wells (see Fig. 4.9), which include two orthogonal spans, the load contribution from the common areas (half landings) may be assumed to be divided equally between both spans.

Other pertinent CP 110 requirements include:

- for staircases spanning in the direction of the flight and which are built into walls (along part or all of their length) a 150 mm strip adjacent to the wall may be deducted from the loaded area (*cf.* Fig. 4.10);
- the effective breadth of a staircase (without stringer beams) should be taken as the actual breadth of the staircase;
- two-thirds of the embedded breadth, up to a maximum of 80 mm should be included in the effective breadth for a staircase that is built into a wall along part or all of its span;
- the depth of the section of a staircase should be taken as the minimum thickness perpendicular to the soffit of the staircase.

A detailed outline of how the actual design is carried out has been given earlier in chapter 4, to which the interested reader should refer.

10.9 Bases

Perhaps the basic difference between the analysis of bases using the working stress methods and that based on the limit state method is the fact that, in the latter case, the determination of moments and forces must be done with ultimate loads. In fact the serviceability limit state of deflection can be ignored for bases, and the bar spacing rules are generally sufficient to control cracks.

The shear consideration should be based on whether direct shear (along a vertical section extending across the full width of the base) or punching shear around the loaded area is more critical.

The procedure outlined in chapter 7 for all types of base is in order, except that ultimate loads must be used instead of service loads.

10.9.1 Local bond

An important component in the design of bases (and also slabs/beams) is the need to prevent local bond failure due to large changes in tension over relatively short lengths of the reinforcement. For the local bond stress f_{bs}, CP 110 recommends the expression:

$$f_{bs} = \frac{V \pm \dfrac{M}{d} \tan \theta_s}{\Sigma U_s d} \qquad (10.8)$$

in which V is the shear force due to ultimate loads; ΣU_s is the sum of the effective perimeters of the tension reinforcement; d is the effective depth of the section; M is the moment at the section due to ultimate loads; and θ_s is the angle between the compression face of the section and the tension reinforcement. If the reinforcing bars are parallel then $\theta_s = 0$ and equation (10.8) degenerates to:

$$f_{bs} = \frac{V}{\Sigma U_s d} \qquad (10.9)$$

Table 10.4 Ultimate local bond stresses (N/mm^2)

Bar type	Concrete grade			
	20	25	30	≥40
Plain bars	1.7	2.0	2.2	2.7
Deformed bars	2.1	2.5	2.8	3.4

The local bond stress f_{bs} computed from the above equations [(10.8) or (10.9)] must not exceed the appropriate value given in Table 10.4. If it does, the most common solution is to increase the number of bars by selecting new bars of smaller sizes (see Example 5.1).

10.10 Exercises

For all the problems listed below assume $f_{cu} = 30\,N/mm^2$ and $f_y = 410\,N/mm^2$.

10.1 Attempt all the problems in section 4.11, using the limit state method.

10.2 Attempt all the problems in section 7.5, using the limit state method.

11 Columns

11.1 Introduction

According to CP 110, a column may be considered short when the ratios l_{ex}/h and l_{ey}/b are less than 12, where l_{ex}, l_{ey} are the effective heights in respect of the major and minor axes, respectively; h is the depth in respect of the major axis; and b is the width of the column. Columns in which l_{ex}/h, $l_{ey}/b > 12$ are considered as slender.

In accordance with CP 110, a column is braced in a given plane if lateral stability to the structure is provided by walls (or some other type of bracing). Otherwise, the column is unbraced (ie, lateral stability is provided by the column itself).

The minimum area of longitudinal reinforcement should normally be at least 1 per cent. The Code allows, for lightly loaded columns, an area of reinforcement A_{sc} such that:

$$A_{sc} \geq 0.15N/f_y \tag{11.1}$$

where N is the ultimate axial load.

The effective column height can be determined using the norms given earlier in Table 5.1. For a framed structure the effective heights l_e may, alternatively, be obtained from:

for a braced column, the lesser of:—

$$l_e = l_0\,[0.7 + 0.05(\alpha_{c1} + \alpha_{c2})] \leq l_0 \tag{11.2}$$

or

$$l_e = l_0\,[0.85 + 0.05\alpha_{cmin}] \leq l_0 \tag{11.3}$$

for an unbraced column, the lesser of:—

$$l_e = l_0\,[1.0 + 0.15(\alpha_{c1} + \alpha_{c2})] \tag{11.4}$$

or

$$l_e = l_0\,[2.0 + 0.3\alpha_{cmin}] \tag{11.5}$$

in which l_0 is the clear height between restraints; α_{c1} is the ratio of the sum of the column stiffnesses to the sum of the beam stiffnesses at one end of the column; α_{c2} is the ratio of the sum of the column stiffnesses to the sum of the beam stiffnesses at the other end of the column; and α_{cmin} is the minimum of α_{c1} and α_{c2}.

A general upper limit of 60 is placed on the slenderness ratio (ie, l_0/least lateral dimension) of a column.

11.2 Axially-loaded short braced columns

For a truly axially-loaded short braced column, at ultimate limit state, the stress in the concrete is $\sigma_c = 0.45f_{cu}$ (Fig. 8.1) and that in the compression steel is $\sigma_{sc} = 0.72f_y$ (Fig. 8.2). Thus, the ultimate axial load would be:

$$N = 0.45f_{cu}A_c + 0.72f_yA_{sc} \tag{11.6}$$

in which A_c is the area of concrete and A_{sc} is the area of the longitudinal reinforcement.

However, in practice there usually exists some eccentricity and to take care of this the ultimate axial load given by equation (11.6) is reduced slightly in CP 110 so that:

$$N = 0.4f_{cu}A_c + 0.67f_yA_{sc} \tag{11.7}$$

For short, braced columns that support an approximately symmetrical arrangement of beams, the ultimate axial load is given by the Code as:

$$N = 0.35f_{cu}A_c + 0.60A_{sc}f_y \tag{11.8}$$

provided that:

- the beams are designed for uniformly-distributed imposed loads; and
- the beam spans do not differ by more than 15 per cent of the longer.

Example 11.1

Determine the ultimate axial load that a 300 mm × 300 mm concrete axially-loaded short, braced column, which is reinforced with four 20 mm ϕ bars, is capable of carrying. Assume $f_{cu} = 30\ \text{N/mm}^2$ and $f_y = 410\ \text{N/mm}^2$.

Solution

$$A_c = 300^2 = 9 \times 10^4\ \text{mm}^2$$

$$A_{sc} = 4 \times \pi \times 20^2/4 = 1257\ \text{mm}^2$$

Using equation (11.7):

$$N = 0.4 \times 30 \times 9 \times 10^4 + 0.67 \times 1257 \times 410$$

$$= \underline{1425\ \text{kN}}$$

11.3 Combined moments and axial forces on short columns

For columns subjected to combined thrusts and moments the requirements of CP 110 are that the minimum design moment should be that caused by the ultimate axial load acting at an eccentricity of $0.05h_{min}$, where h_{min} is the minimum depth of the section. The minimum provision is to cater for eccentricity due to construction tolerances.

11.3.1 Rectangular sections

Now consider a rectangular column section (Fig. 11.1) that is to transmit an ultimate load N and a bending moment M, whose action is such as to cause compression in the upper portion of the section. The equilibrium condition requires that:

$$N = F_c + F_s' + F_s \tag{11.9}$$

where N is the thrust; F_c is the compression in concrete; F_s' is the compression in the reinforcement steel of area A_s ; and F_s is the force [which may be compressive (positive) or tensile (negative)] in the reinforcement steel of area A_s. By setting $F_c = K_1 f_{cu} b_x$, $F_s' = \sigma_s' A_s'$ and $F_s = \sigma_s A_s$, equation (11.9) becomes:

$$N = K_1 f_{cu} b_x + \sigma_s' A_s' + \sigma_s A_s \tag{11.10}$$

in which σ_s' is the compressive stress in the reinforcement of area A_s', while σ_s is the stress (compressive or tensile, depending on the position of the neutral axis) in the reinforcement of area A_s.

By taking moments about the mid-section (the external moment must be balanced by the internal moments formed by the forces F_s', F_c and F_s):

$$M = K_1 f_{cu} b_x (h/2 - K_2 x) + \sigma_s' A_s' (h/2 - d') - \sigma_s A_s (d - h/2) \tag{11.11}$$

in which, according to CP 110, the minimum value of $M = 0.05 h_{min} N$ (where h_{min} is the minimum depth of the section). By defining a characteristic thrust $N_0 = f_{cu} bh$ and moment $M_0 = f_{cu} bh^2$ equations (11.10) and (11.11) can be written in the 'dimensionless' forms:

$$N/N_0 = K_1(x/h) + \rho'(\sigma_s'/f_{cu}) + \rho(\sigma_s/f_{cu}) \tag{11.12}$$

$$M/M_0 = K_1(x/h)[0.5 - K_2(x/h)] + \rho'(\sigma_s'/f_{cu})[0.5 - (d'/h)] + \rho(\sigma_s/f_{cu})[0.5 - (d/h)] \tag{11.13}$$

where $\rho = A_s/bh$ and $\rho' = A_s'/bh$. For a specific value of the neutral axis depth x the strains in the reinforcement (and therefore the stresses σ_s, σ_s') can be determined. Thus, if a symmetrical arrangement of bars is adopted ($\rho = \rho'$) equations (11.12) and (11.13) can be used to obtain dimensionless plots of N/N_0 versus M/M_0 for varying ρ values [assuming fixed values of (x/h), f_{cu} and (d/h)]. Such plots can make the design of columns less complicated and CP 110 (part 2) contains several design charts based essentially on equations (11.10) and (11.11), with $A_s = A_s'$.

In using the above equations to obtain the reinforcement areas A_s and A_s' for given values of M and N, the steps are:

- select a value for the neutral axis depth x;

Fig. 11.1 Rectangular column under combined action of moment and thrust

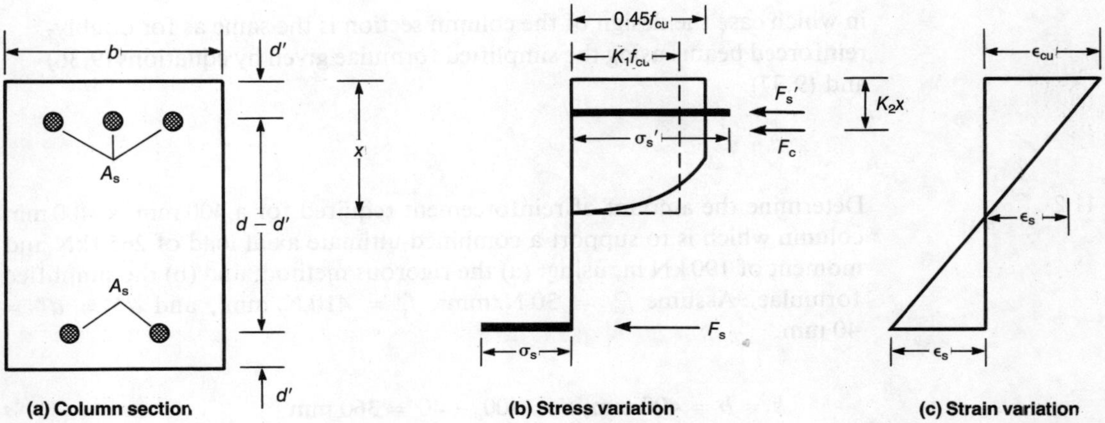

- obtain the strains ϵ_s and ϵ_s' in the reinforcement using the strain diagram [Fig. 11.1(c)];
- determine the stresses σ_s and σ_s' from the stress-strain curve (Fig. 8.2);
- solve equations (11.10) and (11.11) simultaneously for the steel areas A_s and A_s';
- for an optimal design repeat the above steps and read off optimal x from the minimum point on the graph of $(A_s + A_s')$ versus x. Thereafter compute the optimal values of A_s and A_s'.

The Code contains some simplified design formulae for rectangular columns with or without symmetrical arrangement of reinforcement. The formulae consist of three types:

- only nominal reinforcement is required when the ultimate axial load does not exceed

$$N = 0.4 f_{cu} b(h - 2e) \tag{11.14}$$

provided the resultant eccentricity $e = M/N \leq (h/2 - d')$

- when the ultimate axial load is greater than that given by equation (11.14), the strength of the section can be obtained from

$$N = 0.4 f_{cu} b d_c + 0.72 f_y A_s' + \sigma_s A_s \tag{11.15}$$

$$M = 0.2 f_{cu} b d_c (h - d_c) + 0.72 f_y A_s' (h/2 - d') - \sigma_s A_s (d - h/2) \tag{11.16}$$

where d is the depth to the reinforcement A_s (see Fig. 11.1); d_c is the depth of concrete in compression that may be taken as appropriate (although the selection must ensure always that $d_c \geq 2d'$); and σ_s is the stress in reinforcement A_s (which should not exceed $0.72 f_y$ when compressive or $-0.87 f_y$ when tensile). The Code Handbook recommends that:

when $2d' \leq d_c \leq h/2$, $\sigma_s = -0.87 f_y$;
when $h/2 \leq d_c \leq d$, σ_s is to vary linearly from $-0.87 f_y$ to 0;
when $d \leq d_c \leq h$, $\sigma_s = 0$; and
when $d_c = h$, σ_s increases from 0 to $0.72 f_y$.

As an alternative to the preceding method, when $e = M/N \geq (d - h/2)$ the axial load may be ignored and the section designed to resist an increased moment given by:

$$M_a = M + N(d - h/2) \tag{11.17}$$

in which case the design of the column section is the same as for doubly-reinforced beams using the simplified formulae given by equations (9.36) and (9.37).

Example 11.2

Determine the amount of reinforcement required for a 400 mm × 400 mm column which is to support a combined ultimate axial load of 2650 kN and moment of 190 kN m, using: (a) the rigorous method; and (b) the simplified formulae. Assume $f_{cu} = 30\ \text{N/mm}^2$, $f_y = 410\ \text{N/mm}^2$, and $d' = d'' = 40$ mm.

Solution

$h = b = 400$ mm, $d = 400 - 40 = 360$ mm

From equations (9.7) and (9.11):

$$K_1 = 0.403, K_2 = 0.417$$

(a) *The rigorous method*:
Let $x/h = 0.6$, or $x = 240$ mm
From the strain variation [Fig. 11.1(c)]:

$$\epsilon_s = \frac{(d-x)}{x}\epsilon_{cu} = \frac{(360 - 240)}{240} \times 0.0035$$
$$= 0.0018$$

$$\epsilon_s' = \frac{(x - d')}{x}\epsilon_{cu} = \frac{(240 - 40)}{240} \times 0.0035$$
$$= 0.0029$$

From equation (9.13):

$$\epsilon_1 = 410/287\,500 = 0.0014$$

As $\epsilon_s' > 0.002$, from Fig. 8.2:

$$\sigma_s' = \frac{410}{1.15 + 410/2000} = 303 \text{ N/mm}^2$$

and because $\epsilon_1 < \epsilon_s < 0.002$:

$$\sigma_s = \left(\frac{0.0018 - 0.0014}{0.002 - 0.0014}\right)\left(303 - \frac{0.8 \times 410}{1.15}\right) + \frac{0.8 \times 410}{1.15}$$
$$= 289 \text{ N/mm}^2$$

Equations (11.10) and (11.11), with some manipulation, now become:

$$3815 = 1.02A_s' + A_s$$
$$765 = 1.02A_s' - A_s$$

which, when solved, yield:

$$A_s' = 2245 \text{ mm}^2 \text{ and } A_s = 1525 \text{ mm}^2$$

Comments
To obtain an optimal answer, other selections of x can be made and the reinforcement areas A_s', A_s obtained in the same way as above. The value of x that produces the smallest value of $(A_s' + A_s)$ is the best. However, the iteration is left as an exercise to the student.
(b) *Simplified formulae*:
The resultant eccentricity $e = M/N = 190 \times 10^3/2650 = 72$ mm
Using equation (11.14): $N = 0.4 \times 40 \times 400(400 - 2 \times 72)$
$= 1638 \text{ kN} < 2650 \text{ kN}$
Therefore, more than nominal reinforcement is required.
Using equations (11.15) and (11.16), with $d_c = 400$ mm, say, and $\sigma_s = 0.72f_y = 295 \text{ N/mm}^2$:

$$220 = 0.72A_s' - 0.87A_s$$

$$2896 = 0.72A_s' + 0.87A_s$$

which, when solved, yield $A_s' = 2164 \text{ mm}^2$ and $A_s = 1538 \text{ mm}^2$.

Table 11.1 Relationship between N/N_{uz} and factor α_n

N/N_{uz}	α_n
≤ 0.2	1.0
0.4	1.33
0.6	1.67
≥ 0.8	2.0

11.3.2 Non-rectangular sections

The normal procedure, which consists of obtaining the equilibrium equations and then taking moments about a convenient axis, is the first step towards obtaining the pertinent design formulae. Thereafter, the steps outlined earlier in section 11.3.1 should be followed to obtain the appropriate reinforcement areas.

11.4 Short columns in biaxial bending

Where a column section is being considered for biaxial bending, the recommended CP 110 design formulae are:

$$\left(\frac{M_x}{M_{ux}}\right)^{\alpha_n} + \left(\frac{M_y}{M_{uy}}\right)^{\alpha_n} \leq 1.0 \tag{11.18}$$

$$N_{uz} = 0.45 f_{cu} A_c + 0.75 f_y A_s{}' \tag{11.19}$$

where M_x and M_y are the moments about the major and minor axes, respectively, due to ultimate loads; M_{ux} is the maximum moment capacity, assuming ultimate axial load N and bending about the major axis only; M_{uy} is the maximum moment capacity, assuming ultimate axial load N and bending about the minor axis only; and α_n is a factor that depends on the ratio N/N_{uz} as given in Table 11.1. Alternatively, the factors can be obtained from $\alpha_n = {}^5/_3 (N/N_{uz}) + {}^2/_3$ with the limiting conditions $1 \not> \alpha_n \not> 2$.

Example 11.3

A 300 mm × 400 mm concrete column is reinforced with four 25 mm ϕ bars symmetrically arranged as shown in Fig. 11.2. The total cover is 60 mm for all bars.

The column is required to carry an ultimate axial load of 1150 kN along with the moments M_x = 42 kN m, M_y = 25 kN m about the axes x-x and y-y,

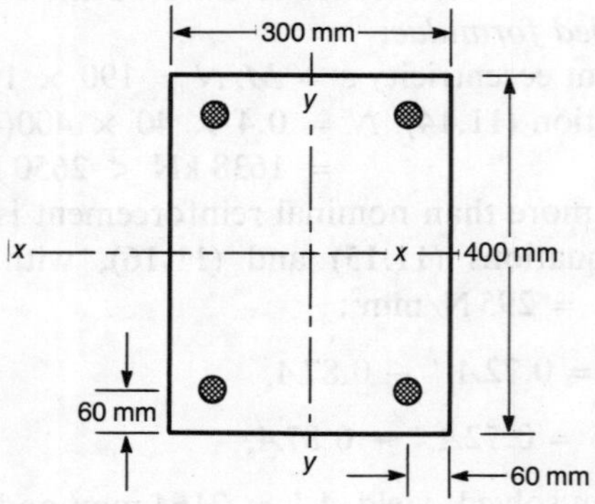

Fig. 11.2 Rectangular column under combined thrust and biaxial bending

respectively. Given that $f_{cu} = 30\ \text{N/mm}^2$ and $f_y = 410\ \text{N/mm}^2$, ensure that the section is adequate for the loading condition.

Solution To compute the maximum moment capacities about the axes, the simplified formula given by equation (11.16) is used:

For M_{ux}, let $d_c = \frac{1}{2}d = \frac{1}{2}(340) = 170$ mm

$$A_s = A_s' = 2 \times \pi \times 25^2/4 = 982\ \text{mm}^2$$

With $\sigma_s' = -0.87f_y$:

$$M_{ux} = 0.2 \times 40 \times 300 \times 170 \times (400 - 170) + 410 \times 982(0.72 - 0.87)(140)$$

or:

$$M_{ux} = 85.38\ \text{kN m}$$

For M_{uy}, let $d_c = \frac{1}{2}(240) = 120$ mm and:

$$M_{uy} = 0.2 \times 40 \times 400 \times 120 \times (300 - 120) + 410 \times 982(0.72 - 0.87)(90)$$
$$= 63.68\ \text{kN m}$$

From equation (11.19):

$$N_{uz} = 0.45 \times 40 \times 300 \times 400 + 0.75 \times 410 \times 928 \times 2$$
$$= 2731\ \text{kN}$$

Therefore:

$$N/N_{uz} = 1150/2731 = 0.42$$

From Table 11.1:

$$\alpha_n = 1.33 + \frac{(0.42 - 0.4)}{(0.6 - 0.4)} \times (1.67 - 1.33)$$
$$= 1.36$$

Therefore:

$$\left(\frac{M_x}{M_{ux}}\right)^{\alpha_n} + \left(\frac{M_y}{M_{uy}}\right)^{\alpha_n} = \left(\frac{42}{85.38}\right)^{1.36} + \left(\frac{25}{63.68}\right)^{1.36}$$
$$= 0.66 < 1.0$$

and the section is capable of resisting the axial thrust and bending moments.

11.5 Slender columns

As expected, slender columns generally develop longer deflections than equivalent short columns under identical loading conditions. Therefore, the design of slender columns must take into account the additional moments induced by the deflection.

Three types of bending can be identified with slender columns. These are when the bending takes place about:

- a minor axis;
- a major axis;

- both minor and major axes.

CP 110 provides appropriate design formulae to be used for each bending condition.

For the first type, when bending occurs about a minor axis, the column should be designed for its ultimate axial load N and a bending moment M_t, given by:

$$M_t = M_i + \frac{Nh}{1750}\left(\frac{l_e}{h}\right)^2\left(1 - 0.0035\frac{l_e}{h}\right) \quad (11.20)$$

in which M_i is the maximum initial moment in the column due to ultimate loads (but not less than $0.05Nh$) calculated using simple elastic analysis; h is the overall depth of the cross-section in the plane of bending; and l_e is the effective height, either in the plane of bending or in the plane at right angles (whichever is greater). When the slender column is braced and there are no transverse loads along its height the Code allows the value of the moment M_i to be reduced to:

$$M_i = 0.4M_1 + 0.6M_2 \quad (11.21)$$

in which M_1 is the smaller initial end moment due to ultimate loads (assumed negative if the column is bent in double curvature); and M_2 is the larger initial end moment due to ultimate loads (assumed positive). For all cases, it is necessary to have $M_t \not< M_2$ and $M_i \not< 0.4M_2$, even if the calculations show otherwise.

A slender column bent about a major axis (the second type of bending described above) can be considered as being biaxially loaded with no initial moment about the minor axis. The Code recommends that, when the overall depth of the section is less than three times the width, a slender column bent about a major axis should be designed for its ultimate axial load N and the moment M_t, given by:

$$M_t = M_i + \frac{Nh}{1750}\left(\frac{l_e}{b}\right)^2\left(1 - 0.0035\frac{l_e}{b}\right) \quad (11.22)$$

where b is the width of the section at right angles to the plane of bending; and M_i, l_e and h are as defined earlier.

For the third type of bending, when the slender column is bent about both axes, the Code requires that the design should be based on the section's ultimate axial load N and moments M_{tx}, M_{ty} about the major and minor axes, respectively. These moments are given by:

$$M_{tx} = M_{ix} + \frac{Nh}{1750}\left(\frac{l_{ex}}{h}\right)^2\left(1 - 0.0035\frac{l_{ex}}{h}\right) \quad (11.23)$$

$$M_{ty} = M_{iy} + \frac{Nb}{1750}\left(\frac{l_{ey}}{b}\right)^2\left(1 - 0.0035\frac{l_{ey}}{b}\right) \quad (11.24)$$

in which M_{ix} is the initial moment due to ultimate loads about the major axis; M_{iy} is the corresponding initial moment about the minor axis; l_{ex} and l_{ey} are the effective heights in the major and minor axes, respectively; h is the depth in respect of the major axis; and b is the width of the column.

The Code allows adjustments to the design moments given by the above formulae, to account for the reduction in lateral deflections for increase in the compression zone of the section. The adjustments can be achieved by multiplying the second terms in equations (11.20) to (11.24) by the factor K given by:

$$K = \frac{N_{uz} - N}{N_{uz} - N_{bal}} \leq 1 \qquad (11.25)$$

in which N is the ultimate axial load; N_{uz} is given in equation (11.19); and N_{bal} is the axial load corresponding to the balanced condition of maximum compressive strain in the concrete (0.35 per cent) and tensile strain in the tension reinforcement (0.2 per cent).

11.6 Exercises

11.1 Calculate the ultimate axial load capacity of a short braced circular (diameter = 400 mm) axially-loaded concrete column that is reinforced with four 25 mm ϕ bars. Assume f_{cu} = 40 N/mm^2 and f_y = 410 N/mm^2.

11.2 Determine, using the rigorous method, the optimal reinforcement amount required for a 300 mm × 300 mm short column which is to support a combined ultimate axial load of 1740 kN and moment of 150 kN m. Assume f_{cu} = 25 N/mm^2, f_y = 410 N/mm^2, with a total cover of 50 mm.

11.3 Attempt Exercise 11.2 again, using the simplified formulae.

11.4 Analyse the column sections in Exercise 5.3 using the limit state method. From the results, produce appropriate charts that can be used to design these sections. Assume the columns are short.

11.5 Attempt Example 11.3 again, given that the value of the ultimate axial load is halved and those of the moments are doubled.

11.6 Obtain a suitable section size and reinforcement for a rectangular column required to carry an axial thrust of 1250 kN along with moments of 260 kN m and 140 kN m about the major and minor principal axes, respectively. Assume f_{cu} = 40 N/mm^2 and f_y = 410 N/mm^2.

11.7 Repeat Exercise 11.6, given that the column section is to be circular.

11.8 A square column is to have a central circular duct of diameter 200 mm. Determine a suitable column size and reinforcement if the section is to carry an axial load of 350 kN and a moment of 460 kN m. Assume f_{cu} = 30 N/mm^2 and f_y = 425 N/mm^2.

11.9 Attempt the previous example again, given that the column is circular and the 200 mm central circular core is to be maintained.

Part Three: **Prestressed concrete, structural detailing and computer-aided design**

12 Prestressed concrete

12.1 Introduction

A structural member is said to be prestressed when an initial ('pre') stress (which is generally of opposite sign to the stresses developed at working load conditions) is introduced into the member. As concrete is weak in tension, one major advantage of a prestressed concrete member over its reinforced concrete counterpart is the ability to ensure that no part of the section is under tension. Other advantages include:

- a reduction in corrosion of the steel because it is possible to eliminate cracks;
- a reduction in the amount of reinforcement required because of reduced tensile stresses;
- an improvement in the shear resistance capacity of sections because of a reduction in the principal tensile stress;
- suitability for mass-production.

The prestressing process necessarily involves the use of high-tensile steel which is used to develop the initial stresses. The method also requires special equipment, such as jacks and anchorages. Prestressing demands a great deal of attention and supervision because of the high quality of concrete required.

12.1.1 Pre-tensioning

In pre-tensioning, the reinforcement wires are tensioned before the concrete is poured. The tension in the wires is then released some days later when the concrete has hardened. The release causes a transfer of stress from the wires into the concrete by means of the bonding action between the steel and concrete.

The importance of bond in pre-tensioning makes the need for small diameter wires quite imperative, because bonding is greatly influenced by the sum total of circumferential lengths. Losses associated with the pre-tensioning process include those due to:

- creep of concrete and steel;
- shrinkage of concrete when stress transfer takes place;
- elastic shortening of concrete;
- friction.

The pre-tensioning process is excellent for precast concrete units that are identical and are to be mass-produced. Such units include poles to carry electricity and concrete floor members. The process allows the use (in a controlled environment such as a factory) of steam-curing methods which can accelerate the hardening of the concrete.

12.1.2 Post-tensioning

In the post-tensioning process, the concrete is cast first in a mould with provision for ducts through which prestressing tendons will subsequently be passed. After the concrete has hardened sufficiently the tendons are tensioned using (usually) a combination of hydraulic jacks and anchorages. The ends of the cables are securely anchored after tensioning.

In post-tensioning, the tendons may be straight or curved, the latter being relatively easy to achieve (when compared with pre-tensioned systems) because the duct has been formed before prestressing. After the tensioning cables have been inserted, the ducts may be grouted under pressure to prevent the corrosion of the wires. Grouting also enhances the strength of the unit. Post-tensioning is highly suitable for in-situ construction, especially of relatively heavy members.

In general, all of the losses associated with pre-tensioning also occur in the post-tensioning system, although some particular losses may be more pronounced (eg, friction due to relative movement between the tendon and the surrounding duct during the tensioning process).

12.1.3 Characteristic strengths of concrete and prestressing tendons

The characteristic strengths for different grades of concrete specified by CP 110 are quoted in Table 12.1, along with the related cube strengths at other ages. The prestressed concrete design can be based either on the characteristic strength or, when appropriate, on the strength given in Table 12.1 for the appropriate age of loading.

For the prestressing tendons (wires, strands and bars) the specified characteristic strengths, as given in CP 110, are quoted in Tables 12.2 to 12.4.

12.2 Analysis of sections under working load conditions

In this section, the relationship between the prestressing force P and the stresses developed across the cross-section of the prestressed concrete member, under working load or service conditions, is examined. The goal is to obtain the appropriate conditions for which the stresses in the concrete range in value between a prescribed maximum compressive stress σ_{max} and a minimum stress σ_{min} [which may be compressive (positive) or tensile (negative)]. In the majority of prestressed concrete problems, it is desirable to have $\sigma_{min} \geq 0$, so as to prevent the development of cracks in the section.

The analysis is based on the following assumptions:

- plane sections always remain plane, so that the strain variation across the section is linear;
- there is a linear relationship between stress and strain;
- the initial prestressing force P (at transfer) has been reduced (due to losses) to ηP at working load conditions. [This assumption differs somewhat from the one given by CP 110 (clause 4.3.3.1) in which section analysis is based on stresses developed after the transfer of prestress and

Grade	Characteristic strength f_{cu} (N/mm²)	Cube strength (N/mm²) at age of				
		7 days	2 months	3 months	6 months	1 year
30	30.0	20	33	35	36	37
40	40.0	28	44	45.5	47.5	50
50	50.0	36	54	55.5	57.5	60
60	60.0	45	64*	65.5*	67.5*	70*

*These increased strengths due to age should only be used if it has been demonstrated to the satisfaction of the engineer that the materials to be used are capable of producing these higher strengths.

Table 12.1 Strength of concrete

Nominal size (mm)	Specified characteristic strength $A_{ps}f_{pu}$ (kN)	Nominal cross-sectional area A_{ps} (mm²)
2	6.34	3.14
2.65	10.3	5.5
3	12.2	7.1
3.25	14.3	8.3
4	21.7	12.6
4.5	25.7	15.9
5	30.8	19.6
7	60.4	38.5

Table 12.2 Specified characteristic strengths of prestressing wire

Number of wires	Nominal size (mm)	Specified characteristic strength $A_{ps}f_{cu}$ (kN)	Nominal cross-sectional area A_{ps} (mm²)
7	6.4	44.5	24.5
	7.9	69.0	37.4
	9.3	93.5	52.3
	10.9	125	71.0
	12.5	165	94.2
	15.2	227	138.7
19	18	370	210
	25.4	659	423
	28.6	823	535
	31.8	979	660

Table 12.3 Specified characteristic strengths of prestressing strand

Nominal size (mm)	Specified characteristic strength $A_{ps}f_{cu}$ (kN)	Nominal cross-sectional area A_{ps} (mm²)
*20	325	314
22	375	380
*25	500	491
28	625	615
*32	800	804
35	950	961
*40	1250	1257

*preferred sizes.

Table 12.4 Specified characteristic strengths of prestressing bars

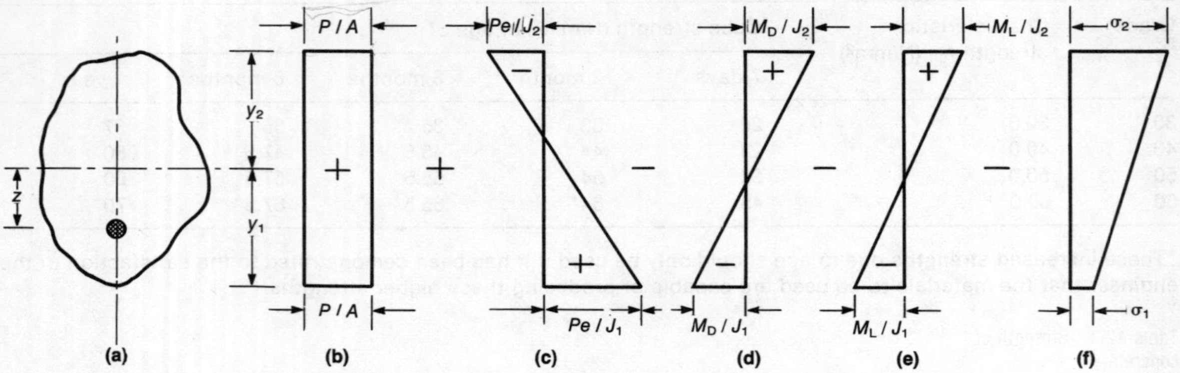

Fig. 12.1 Prestressed section (arbitrary shape)

after all prestress losses have occured. To base the analysis on such an assumption, it is sufficient to take P as the prestressing force after all losses (ie, ηP above) and thereafter set $\eta = 1$ in the various expressions derived below];

- bending takes place about a principal axis.

Now consider the prestressed concrete section of arbitrary shape shown in Fig. 12.1(a). Let the prestressing force P be applied at an eccentricity e from the centroid of the section. The bending moments at the section due to dead and live loads are M_D and M_L, respectively.

As established in chapter 5 for members under combined actions of thrusts and moments, the stress variations across the arbitrarily-shaped section are as shown in Figs. 12.1(b) to (e) where (if J is the section modulus and I the second moment of area): $J_1 = I/y_1$ and $J_2 = I/y_2$. Fig. 12.1(f) represents the effect of the prestressing force, combined with those of the dead and live loads.

Two separate conditions must be identified:

- an initial prestressing state, when the force is P and only the dead load is to be considered;
- a working load state, when the force has reduced to ηP and both the dead and live loads must be taken into account.

Under the first condition, there will be a tendency for larger compressive stresses to develop in the lower portion of the section than in the upper part. It is therefore necessary to have:

$$\sigma_1 = P/A + Pe/J_1 - M_D/J_1 \leq \sigma_{max} \quad (12.1)$$

and

$$\sigma_2 = P/A - Pe/J_2 - M_D/J_2 \geq \sigma_{min} \quad (12.2)$$

Under working load conditions, larger stresses will develop in the upper portion of the section and the following must be enforced:

$$\sigma_1 = \eta P/A + \eta Pe/J_1 - (M_D + M_L)/J_1 \geq \sigma_{min} \quad (12.3)$$

$$\sigma_2 = \eta P/A - \eta Pe/J_2 + (M_D + M_L)/J_2 \leq \sigma_{max} \quad (12.4)$$

Equations (12.1) to (12.4) are four expressions which can be manipulated in order to analyse or design a prestressed concrete section. When the prestressing force is purely axial, then $e = 0$ above, and if the section is just in compression under working load conditions, it can easily be shown that the largest compressive stress developed is:

$$\sigma_2 = \eta(1 + J_1/J_2)P/A \tag{12.5}$$

Going back to the more general case of eccentric prestressing force, the limiting conditions can be obtained by changing the inequalities in equations (12.1) to (12.4) to equalities. Then, by subtracting equation (12.1) from equation (12.3), and equation (12.2) from equation (12.4):

$$(1-\eta)P/A + (1-\eta)Pe/J_1 + M_L/J_1 = (\sigma_{max} - \sigma_{min}) \tag{12.6}$$

$$(1-\eta)P/A - (1-\eta)Pe/J_2 - M_L/J_2 = (\sigma_{min} - \sigma_{max}) \tag{12.7}$$

which can be rearranged to give:

$$J_1 = \frac{(1-\eta)Pe + M_L}{(\sigma_{max} - \sigma_{min}) - (1-\eta)P/A} \tag{12.8}$$

and:

$$J_2 = \frac{(1-\eta)Pe + M_L}{(\sigma_{max} - \sigma_{min}) + (1-\eta)P/A} \tag{12.9}$$

To obtain a trial section for design simply set $\eta = 1$ (ie, no loss of prestress) in the above equations (12.8) and (12.9) and obtain:

$$J_1 = J_2 = J = M_L/(\sigma_{max} - \sigma_{min}) \tag{12.10}$$

For example, for a rectangular section of width b and overall depth h, $J_1 = J_2 = {}^1\!/_{12}bh^3/{}^1\!/_2h = {}^1\!/_6bh^2$ and the trial section can be selected such that:

$${}^1\!/_6bh^2 = M_L/(\sigma_{max} - \sigma_{min}) \tag{12.11}$$

or, if b is prescribed:

$$h = \sqrt{\{6M_L/b(\sigma_{max} - \sigma_{min})\}} \tag{12.12}$$

In the limiting condition the two most critical equations are equations (12.2) and (12.3), because of the absolute necessity to ensure that no tensile stresses develop. By eliminating P between these equations, with $J_1 = J_2 = J$, the limiting eccentricity e_0 can be obtained as:

$$e_0 = \frac{J/A[(1-\eta)J\,\sigma_{min} + (1+\eta)M_D + M_L]}{[(1+\eta)\sigma_{min} + (1+\eta)M_D + M_L]} \tag{12.13}$$

which can be used as a first estimate of the required eccentricity.

The limiting prestressing force can also be shown to be:

$$P_0 = \frac{\sigma_{min}\,J - M_D}{J/A - e_0} \tag{12.14}$$

If, in particular, $\sigma_{min} = 0$, then:

$$e_0 = \frac{J/A[(1+\eta)M_D + M_L]}{[(1-\eta)M_D + M_L]} \tag{12.15}$$

and:

$$P_0 = \frac{M_D}{e_0 - J/A} \tag{12.16}$$

Example 12.1 A 300 mm wide rectangular prestressed concrete beam of span 12 m is required to carry a uniformly-distributed live load of 10 kN/m. A loss of 15 per cent in the prestressing force is expected under working load conditions. The permissible compressive stress in concrete is 16 N/mm² and no tensile stresses are allowed. Given that the prestressing wires can be stressed to 950 N/mm², determine:

(a) the overall depth of section required; and
(b) the appropriate number of 7 mm diameter wires.

Solution The bending moment at mid-span (which is the most critical section) due to the live load is:

$$M_L = 10 \times 12^2/8 = 180\,\text{kN m}$$

Using equation (12.12), with $\sigma_{max} = 16\,\text{N/mm}^2$ and $\sigma_{min} = 0$, the trial depth is:

$$h = \sqrt{\left(\frac{6 \times 180 \times 10^6}{300 \times 16}\right)} = 474\,\text{mm}$$

Therefore, try a section 300 mm × 480 mm.

The weight (assuming 23.6 kN/m³) of the beam = 0.3 × 0.48 × 23.6 = 3.4 kN/m. Therefore, the dead load moment M_D is given by:

$$M_D = 3.4 \times 12^2/8 = 61.2\,\text{kN m}$$

Using equation (12.15), with $J/A = \frac{1}{6}bh^2/bh = h/6$:

$$e_0 = \frac{480}{6}\left[\frac{(1 + 0.85) \times 61.2 + 180}{(1 - 0.85) \times 61.2 + 180}\right]$$

$$= 124\,\text{mm}$$

which, when introduced into equation (12.16), yields:

$$P_0 = \frac{61.2 \times 10^3}{124 - 480/6} = 1391\,\text{kN}$$

A check should now be made to see whether equations (12.1) and (12.3) have been violated. Using equation (12.1):

$$\sigma_1 = 1391/480 \times 300 + \frac{(1391 \times 0.124 - 61.2) \times 10^3}{\frac{1}{6} \times 300 \times 480^2}$$

$$= 19.3\,\text{N/mm}^2 > 16\,\text{N/mm}^2$$

and using equation (12.3):

$$\sigma_2 = 0.85 \times 1391/480 \times 300 - \frac{(0.85 \times 1391 \times 0.124 - 61.2 - 180) \times 10^3}{\frac{1}{6} \times 300 \times 480^2}$$

$$= 16.4\,\text{N/mm}^2 > 16\,\text{N/mm}^2.$$

Comments

The above solution to Example 12.1 shows that, at transfer, the maximum compressive stress in concrete (σ_1) exceeds the permissible stress by about 21 per cent, while at working load the difference is 2.5 per cent. The requirements of CP 110 allow the compressive stresses at transfer to be up to 50 per cent of the concrete strength at transfer (when the prestress distribution is triangular or near triangular), or 40 per cent of the same (when the distribution is uniform). As the concrete used in Example 12.1 is essentially of grade 50 (as may be inferred from the allowable compressive stress) the computed maximum stress at transfer may still be acceptable (see Table 12.1 for the cube strengths at different ages).

Furthermore, the rule of thumb is that, at working load, the maximum stress σ_2 may exceed the permissible stress by up to 10 per cent as σ_2 is compressive in any case. Therefore, the computed stresses are not 'too unacceptable'.

However, for a more conservative design the overall depth can be increased to 520 mm, so that the dead load = 0.3 × 0.52 × 23.6 = 3.68 kN/m and:

$$M_D = 3.68 \times 12^2/8 = 66.3 \text{ kN m}$$

$$e_0 = \frac{520}{6}\left[\frac{1.85 \times 66.3 + 180}{0.15 \times 66.3 + 180}\right] = 138.1 \text{ mm}$$

$$P_0 = \frac{66.3 \times 10^3}{138.1 - 520/6} = 1289 \text{ kN}$$

The maximum stress at transfer is now:

$$\sigma_1 = 1289/520 \times 300 + \frac{(1289 \times 0.1381 - 66.3) \times 10^3}{\frac{1}{6} \times 300 \times 520^2} = 16.5 \text{ N/mm}^2$$

(which is now only slightly larger than 16.0 N/mm^2), while at working load the maximum is:

$$\sigma_2 = 0.85 \times 1289/520 \times 300 - \frac{(0.85 \times 1289 \times 0.1381 - 66.3 - 180) \times 10^3}{\frac{1}{6} \times 300 \times 520^2} = 14.0 \text{ N/mm}^2 < 16.0 \text{ N/mm}^2.$$

The number of 7 mm diameter wires required is:

$$\frac{1289 \times 10^3}{\pi/4 \times 7^2 \times 950} = 35$$

Equations (12.1) to (12.4) can also be used to design the appropriate tendon profile (represented by the eccentricity e) that ensures that the stresses developed at all sections (not only the critical one) are within the allowable values. The equations can be rewritten (in the form of constraint expressions for the eccentricity of the prestressing force) as:

$$e \leq J_1/P[\sigma_{max} - P/A] + M_D/P \quad (12.17)$$

$$e \leq J_2/P[P/A - \sigma_{min}] + M_D/P \quad (12.18)$$

$$e \geq J_1/\eta P[\sigma_{min} - \eta P/A] + (M_D + M_L)/\eta P \quad (12.19)$$

$$e \geq J_2/\eta P[\eta P/A - \sigma_{max}] + (M_D + M_L)/\eta P \quad (12.20)$$

For a given prismatic member and prestressing force, the first terms in equations (12.17) to (12.20) are essentially constants. When the loading is uniformly distributed, the moments M_D and M_A have a parabolic distribution along the length of the member. Thus, the constraint equations suggest that the tendon profile must be parabolic if a straight profile will not fall within the allowable cable zone ($e_{max} - e_{min}$), where e_{max} is given by the smaller of equations (12.17) and (12.18) and e_{min} is the larger of equations (12.19) and (12.20). For a more general (but not necessarily uniform) loading situation the allowable cable zone is determined by the shapes of the bending moment diagrams.

12.3 Analysis at ultimate limit state

The flexural analysis of a prestressed concrete section under ultimate loads is essentially based on the same assumptions as those used for reinforced concrete sections, with the additional assumptions that (CP 110, clause 4.3.4.1):

12.3.1 Flexure

- the strains in bonded prestressing tendons and in any additional reinforcement (whether in tension or compression) are derived from the assumption that plane sections remain plane;
- the stresses in bonded prestressing tendons (whether initially tensioned or untensioned) and in additional reinforcement, are derived from the appropriate stress-strain curves given in Figs. 12.2 and 12.3 for tendons and Fig. 8.2 for reinforcement;
- in post-tensioned members where the tendons are unbonded, the stress in the tendons does not exceed the values given in Table 12.5, unless a higher stress can be justified on the basis of tests.

Given these assumptions, the ultimate limit state analysis can be carried out using the methods described in section 9.1.

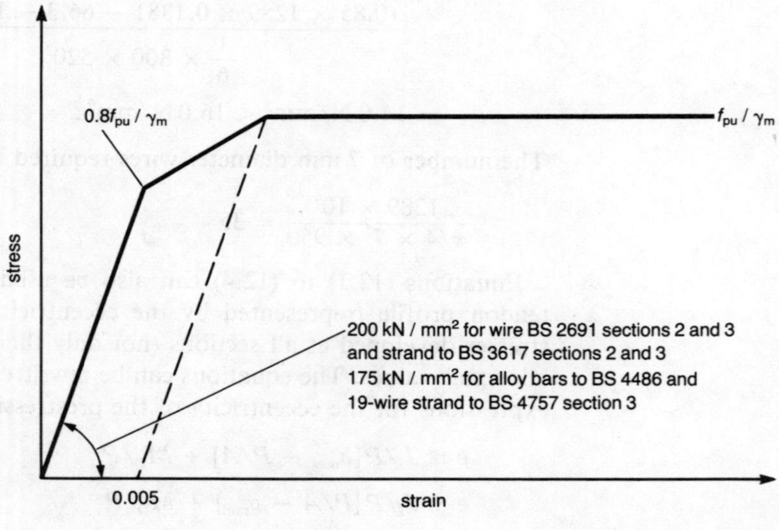

Fig. 12.2 Short-term design stress-strain curve for normal and low relaxation products

Fig. 12.3 Design stress-strain curve for 'as drawn' wire and 'as spun' strand

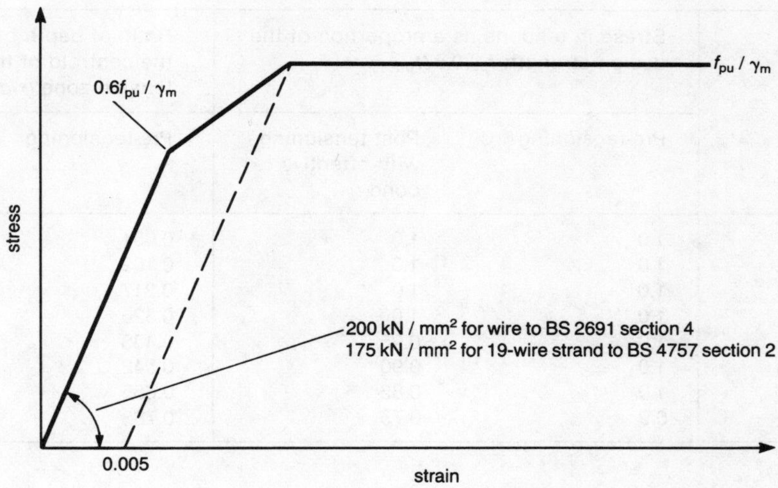

$\frac{f_{pu} A_{ps}}{f_{cu} bd}$	Stress in tendons as a proportion of effective prestress f_{pb}/f_{ps} for value of: $\frac{1}{d}\left(\frac{\text{effective span}}{\text{effective depth}}\right)$			Ratio of depth of neutral axis to that of the centroid of the tendons in the tension zone x/d for values of: $\frac{1}{d}\left(\frac{\text{effective span}}{\text{effective depth}}\right)$		
	30	20	10	30	20	10
0.0025	1.23	1.34	1.45	0.10	0.10	0.10
0.05	1.21	1.32	1.45	0.16	0.16	0.18
0.10	1.18	1.26	1.45	0.30	0.32	0.36
0.15	1.14	1.20	1.36	0.44	0.46	0.52
0.20	1.11	1.16	1.27	0.56	0.58	0.64

Table 12.5 Conditions at the ultimate limit state for post-tensioned rectangular beams having unbonded tendons

The design formula recommended by CP 110 (in the absence of a more fundamental analysis based on the preceding assumptions) for a rectangular beam is:

$$M_u = f_{pb} A_{ps} (d - 0.5x) \tag{12.21}$$

in which M_u is the moment of resistance of the section; f_{pb} is the tensile stress in the tendons at failure; A_{ps} is the area of the prestressing tendons; and x and d are as described earlier.

Table 12.6 contains values of f_{pb} and x for both pre-tensioned members and post-tensioned members with an effective bond between the concrete and tendons.

The allowable effective prestress after all losses should be at least $0.45f_{pu}$. When the design formula is being used for strength calculations, the prestressing tendons and additional reinforcement in the compression zone are not to be taken into account.

12.3.2 Shear

In calculating the ultimate shear resistance V_c of prestressed beams and slabs, either V_{co} or V_{cr} (whichever is the lesser) is used, where V_{co} is the shear

$\frac{f_{pu} A_{ps}}{f_{cu} bd}$	Stress in tendons as a proportion of the design strength $f_{pb}/0.87f_{pu}$		Ratio of depth of neutral axis to that of the centroid of the tendons in the tension zone (x/d)	
	Pre-tensioning	Post-tensioning with effective bond	Pre-tensioning	Post-tensioning with effective bond
0.025	1.0	1.0	0.054	0.054
0.05	1.0	1.0	0.109	0.109
0.10	1.0	1.0	0.217	0.217
0.15	1.0	1.0	0.326	0.326
0.20	1.0	0.95	0.435	0.414
0.25	1.0	0.90	0.542	0.488
0.30	1.0	0.85	0.655	0.558
0.40	0.9	0.75	0.783	0.653

Table 12.6 Conditions at the ultimate limit state for rectangular beams with pre-tensioned tendons or with post-tensioned tendons having effective bond

resistance of a section uncracked in flexure and V_{cr} is that of a section cracked in flexure. This method must be followed, whether or not the section is assumed to be cracked at the onset (under working load conditions), because a prestressed concrete member which is uncracked under working loads may be cracked at some sections when carrying ultimate loads.

To obtain V_{co}, a maximum principal stress f_t ($= 0.24\sqrt{f_{cu}}$) at the centroidal axis of the section is used and the ultimate shear resistance is given by:

$$V_{co} = 0.67bh\sqrt{(f_t^2 + 0.8f_{cp}f_t)} \qquad (12.22)$$

where f_{cp} is the compressive stress at the centroidal axis due to prestress; b is the breadth of the member (which for T-, I- and L-beams should be replaced by the breadth of the rib b_w); and h is the overall depth of the member.

For a section that is cracked in flexure the ultimate shear resistance formula given by CP 110 is:

$$V_{cr} = \left(1 - 0.55\frac{f_{pe}}{f_{pu}}\right) v_c\, bd + M_0V/M \qquad (12.23)$$

where d is the distance from the extreme compression fibre to the centroid of the tendons at the section considered; M_0 ($= 0.8f_{pe}\, I/y$) is the moment necessary to produce zero stress in the concrete at depth d and distance y from the centroid of the concrete section (the second moment of area of which is I); f_{pe} is the effective prestress after all losses have occurred; v_c is the ultimate shear stress, which can be obtained from Table 9.8; and V and M are the shear force and bending moment, respectively, at the section due to ultimate loads. The minimum value for V_{cr} recommended in the Code is $0.1\, bd\sqrt{f_{cu}}$.

In prestressed concrete slabs, shear reinforcement need not be provided whenever $V < V_c$. However, in beams for which $V < V_c$ shear reinforcement need not be provided when:

- $V < 0.5V_c$;
- the members are of minor importance;

- tests have shown shear reinforcement is unnecessary.

The recommendations of CP 110 are that minimum shear reinforcement (in the form of links) be provided in all other members, such that the spacing s_v of the link is related to the cross-sectional area A_{sv} of the two legs of a link by the expression:

$$\frac{A_{sv}}{s_v} = \frac{0.4b}{0.87f_{yv}} \tag{12.24}$$

where f_{yv} is the characteristic strength of the reinforcement, which shall not be taken to be greater than 425 N/mm². In cases where $V > V_c$ the shear reinforcement should, in addition, be such that:

$$\frac{A_{sv}}{s_v} = \frac{(V - V_c)}{0.87f_{yv}d_t} \tag{12.25}$$

where d_t is the depth from the extreme compression fibre to the longitudinal reinforcement bars, or to the centroid of the tendons, whichever is greater. The spacing of the links should not exceed $0.75d_t$ nor four times the web thickness for flanged members. When $V > 1.8V_c$, the maximum spacing should be reduced to $0.5d_t$. The lateral spacing of the individual legs of the links at a section should not exceed $0.75d_t$.

Table 12.7 shows the maximum shear stress values recommended by CP 110. The shear force V, due to ultimate loads, should not exceed the appropriate value of the maximum stress multiplied by bd.

Table 12.7 Maximum shear stress

	Concrete grade			
	30	40	50	≥60
Maximum shear stress (N/mm²)	4.1	4.7	5.3	5.8

12.4 Serviceability limit states

Prestressed concrete members are divided into three classes in CP 110 and these are:

Class 1: No tensile stresses are allowed under service (or working) load conditions;

Class 2: Tensile stresses are allowed, but these must be such that there exists no visible cracking; and

Class 3: There are tensile stresses and cracks but the surface widths of the cracks do not exceed 0.1 mm for members exposed to an aggressive environment (see **severe** category in Table 9.6).

The allowable compressive stresses in the concrete for the serviceability limit states are given in Table 12.8.

For class 2 members, the tensile stresses should not be greater than the design flexural tensile strength of the concrete for pre-tensioned members, nor 0.8 of the design flexural tensile strength for post-tensioned members. The allowable tensile stresses are quoted in Table 12.9.

The Code allows the stress given by Table 12.9 to be increased by up to 1.7 N/mm², provided that tests have shown that such enhanced stress does not exceed 75 per cent of the tensile stress (calculated from the loading in

Table 12.8 Compressive stresses in concrete for serviceability limit states

Nature of loading	Allowable compressive stresses
Design load in bending	$0.33f_{cu}$
	In continuous beams and other statically indeterminate structures this may be increased to $0.4f_{cu}$ within the range of support moments
Design load in direct compression	$0.25f_{cu}$

Table 12.9 Allowable flexural tensile stresses for class 2 members: serviceability limit state of cracking

	Allowable stress (N/mm²) for concrete grade			
	30	40	50	60
Pre-tensioned members	—	2.9	3.2	3.5
Post-tensioned members	2.1	2.3	2.55	2.8

Table 12.10 Hypothetical flexural tensile stresses for class 3 members

Group	Limiting crack width (mm)	Stress (N/mm²) for concrete grade		
		30	40	≥50
A. Pre-tensioned tendons	0.1	—	4.1	4.8
	0.2	—	5.0	5.8
B. Grouted post-tensioned tendons	0.1	3.2	4.1	4.8
	0.2	3.8	5.0	5.8
C. Pre-tensioned tendons distributed in the tensile zone and positioned close to the tension faces of the concrete	0.1	—	5.3	6.3
	0.2	—	6.3	7.3

Table 12.11 Depth factors for tensile stresses for class 3 members

Depth (mm) of member				
≤200	400	600	800	≥1000
1.1	1.0	0.9	0.8	0.7

the performance test corresponding to the appearance of the first crack). Where such an increase is used, the stress in the concrete due to prestress (after losses) should be at least 10 N/mm²; pre-tensioned tendons should be well distributed throughout the tension zone of the section and post-tensioned tendons should be supplemented, if necessary, by additional reinforcement located near the tension face of the member.

For class 3 members, it may be assumed that the concrete section is uncracked and that hypothetical tensile stresses exist at the maximum size of cracks defined earlier. The hypothetical tensile stresses that can be used for members with either pre-tensioned or grouted post-tensioned tendons are shown in Table 12.10.

Because the cracking in a prestressed concrete member subjected to bending is influenced by the overall depth, CP 110 has given some modification factors which should be used in multiplying the hypothetical tensile stresses given in Table 12.10. These modification or depth factors are quoted in Table 12.11.

12.5 Exercises

12.1 A 250 mm wide rectangular prestressed concrete beam of span 10 m is required to carry a uniformly-distributed live load of 12.5 kN/m. A loss of 10 per cent in the prestressing force is expected at working load conditions. The permissible compressive stress in concrete is 14 N/mm^2; the beam is a class 1 member. Given that the prestressing wires can be stressed to 800 N/mm^2, determine:

- the overall depth of section required; and
- the number of 5 mm diameter wires.

12.2 Attempt Exercise 12.1 again, assuming the beam is a class 2 pre-tensioned member and the concrete is of grade 40.

12.3 A 280 mm × 280 mm prestressed concrete beam is simply supported over a span of 7.5 m. The live load consists of a uniformly-distributed load of 8.6 kN/m. The prestressing tendon is straight and is located at an eccentricity e = 72 mm below the centroid of the section. Given that the beam is a class 1 member, obtain the appropriate prestressing force required. Assume f_{cu} = 50 N/mm^2.

12.4 If the beam in Exercise 12.3 were to be replaced by one with a trapezoidal section (base width = 200 mm, top width 350 mm, overall depth = 280 mm), with the same span and loading condition, obtain the minimum prestressing force required.

12.5 Determine the ultimate resistance moment of a 200 mm × 250 mm pre-tensioned concrete beam which is stressed by six 7 mm diameter wires of characteristic strength f_{pu} = 1570 N/mm^2. The wires are at an eccentricity 60 mm from the centroid of the section. The initial prestressing force is 950 kN and the losses have been estimated to be 20 per cent. Assume f_{cu} = 50 N/mm^2.

13 Structural detailing

13.1 Introduction

Structural detailing is basically an art by which the design of a structural system is 'draughted' in a form that the builder/contractor can use for construction. Structural detailing is undoubtedly a crucial aspect of the design process. There is no gainsaying the fact that a sound design can be marred by bad structural details.

A detail drawing is, perhaps, the singularly most important link between the design engineer and the contractor. A good detail drawing can save several man-hours on the construction site. On the other hand a bad detail drawing, apart from making wasteful demands on the site engineer's time and energy, may often create confusion and animosity between the consultants and the contractor.

A structural detail drawing is considered good when it is functional and not just simply aesthetically appealing. It must be easy to read, and details of important areas of the structure must be clearly shown where necessary. Dimensions must be clear, and extensive (but not verbose) notes giving relevant information (eg, expected strength or mix of concrete and the reinforcement type assumed in the design) that cannot otherwise be shown in the drawings, should be provided.

Relatively smaller drawing sheets called the bar bending schedules (or BBS, for short) showing details of how the reinforcement bars are to be bent, should accompany the main drawings.

The BBS, together with the main detail drawings, are essential to the quantity surveyor, who has the responsibility of preparing the so-called bill of quantities used in assessing the cost of the entire project. The BBS is also useful for the steel supplier (who may be the contractor) and the 'iron-bender'.

This chapter contains an outline of the important components of the structural detailing process. In subsequent sections typical detail drawings of the major structural members (ie, slabs, beams, columns and bases) which were dealt with in earlier chapters, are presented.

It is worth emphasising, however, that structural detailing is best learnt by actually carrying out the drawings and not by merely reading about it. The aim here is to make the process of starting less of a problem.

Some engineers think (erroneously) that the preparation of structural details should not be learnt (and mastered) by the designer but should be left to draughtsmen. However, the designer should be proficient in structural detailing so that he may, at least, be able to check that his design has been adequately represented in the drawing (even when prepared by a draughtsman). The need for the engineer designer to have a working knowledge of structural detailing cannot be over-emphasised.

13.2 General guidelines

An important aspect of the preparation of structural details is the choice of scales. For the **general layout**, which shows the location of the various structural elements (eg, beams, slabs and columns) on a single sheet, a scale of 1:50 or 1:100 is the most common. Suitable scales for slabs are 1:50 (for the commonest types) and 1:20 (for more complicated types). When sections of beams and columns are shown, the usual scales are 1:10 or 1:20. However when these elements are shown in elevation a more appropriate scale would be 1:20.

Although styles vary from one design office to another, bar mark numbers should not be complicated and should convey relevant information such as the bar size and possibly the structural member of which it is a part. Each structural member should be given a name which the reader could easily use to identify the location of the unit within the entire structure (eg, FFB-3 for a first floor beam — number 3). When bars are 'called up', the bar indicator must contain important information such as: the bar size; bar mark; reinforcement type; number required (if necessary); spacing (when required); placement (top or bottom); and any other relevant information (eg, arrangement of bars).

These general guidelines have been followed in the preparation of detail drawings of some structural elements shown in section 13.4.

13.3 Curtailment, anchorage and lap lengths of reinforcement

Principally, because of the need for a secure grip (bond) between concrete and reinforcement, design codes provide specific rules formulated to ensure that the bars are properly curtailed and anchored. Where two separate bars overlap, the lap lengths must be such that the bars actually behave as one continuous reinforcement. The curtailment, anchorages and laps must be well represented in the detail drawings.

The general requirements of NCP 1 and CP 110, in respect of anchors and laps in bars, are presented below.

13.3.1 NCP 1 requirements

It is convenient to summarise the requirements as follows (quoted largely from section 3.11 NCP 1):

- a bar in tension should extend from any section for a distance to the end of the bar such that the average bond stress is within permissible values (see Table 2.2). To satisfy this condition, the length of the bar measured from such a section should not be less than:

$$l_e = \phi\sigma_{st}/4\sigma_{bo} \quad (13.1)$$

in which ϕ is the bar diameter, σ_{st} is the tensile stress in the bar, and σ_{bo} is the permissible average bond stress. The bar should also extend at least

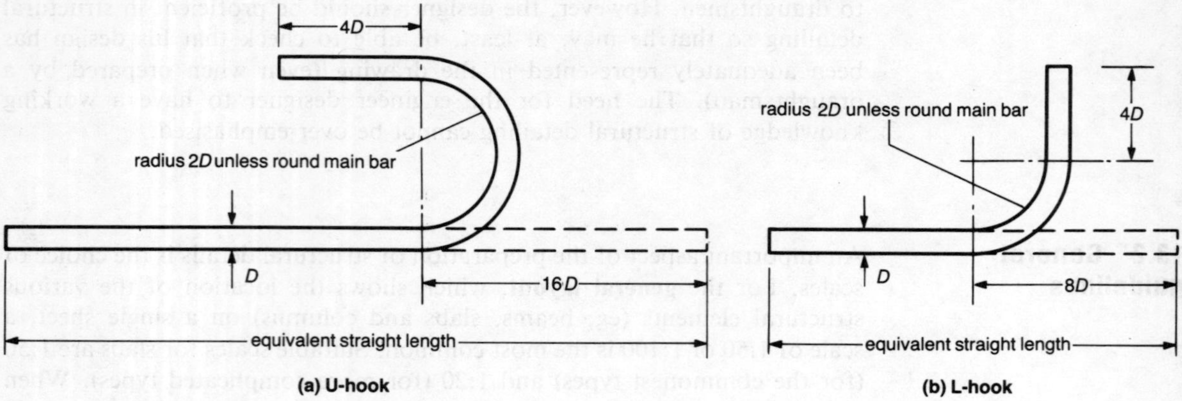

Fig. 13.1 Types of hook

12 bar diameters beyond the point at which it is no longer required to resist stress;

- hooks, when used, are to be of the U-type or L-type (Fig. 13.1) although the U-type is to be preferred. When mild steel bars are used, the hooks must be such that —
  (i) the internal radius of the bend is at least equal to 2ϕ, except where the hook fits over a main reinforcing or other adequate anchor bar, in which case the radius of the bend may be reduced to that of such a bar,
  (ii) the length of a straight bar beyond the end of the curve is not to be less than 4ϕ. When the hooks consist of high-yield bars, the internal radius of the bend should not be less than 3ϕ and the length of the straight bar beyond the end of the curve should be at least 4ϕ;
- the anchorage value of a bend is equivalent to a length of bar equal to 4ϕ for each 45° through which the bar is bent, provided that —
  (i) the radius of the bend is at least equal to 2ϕ,
  (ii) the length of the straight part of the bar beyond the end of the curve is at least 4ϕ,
  (iii) the assumed anchorage value is not taken as more than that equivalent to a length of bar equal to 16ϕ, irrespective of the angle through which the bar is bent.
- laps in bars in any member should be staggered. The length of lap in bars in tension should not be less than that given by equation (13.1) or 30ϕ, whichever is the greater. For bars in compression, the lap length should not be less than:

$$l_p = \phi\, \sigma_{sc}/5\sigma_{bo} \qquad (13.2)$$

or 24ϕ, whichever is the greater (σ_{sc} is the compressive stress in the bar). Other NCP 1 requirements relevant to the preparation of detail drawings have been given elsewhere within chapters 3 to 7.

13.3.2 CP 110 requirements

The British Code of Practice CP 110 also recommends the use of U- and L-type hooks as reinforcement anchorages. The Code (clause 3.11.6.7) requires that the effective anchorage length of a hook or bend should be measured from the start of the bend to a point 4ϕ beyond the end of the bend, and may be taken as the lesser of 24ϕ or:

- *for a hook*, eight times the internal radius of the hook;
- *for a 90° bend*, four times the internal radius of the bend.

Other relevant requirements of the Code in respect of curtailment and anchorage of reinforcement include the following:

- in members under flexure, every bar must extend (except at end supports) beyond the point at which it is no longer needed for a distance equal to d (the effective depth) or 12ϕ, whichever is greater. The point at which reinforcement is no longer required is obviously where the bending moment is exactly equal to the section resistance moment (considering only continuing bars);
- at the simply-supported end of a member, each tension bar should be anchored by one of the following:
  (i) an effective anchorage equivalent to 12ϕ beyond the centreline of the support (no bend or hook should begin before the centre of the support),
  (ii) an effective anchorage equivalent to $12\phi + d/2$ from the face of the support (no bend should begin before $d/2$ from the face of the support),
  (iii) provided the local bond stress at the face of a support is less than the value given by equation (10.8), a straight length of bar beyond the centreline of the support equal to either one third the support width or 30 mm, whichever is the greater.

Simplified rules (as alternatives to the above) that can be used for curtailment of bars in beams and slabs which support essentially uniformly-distributed loads, are provided by CP 110. These simplified rules are given below.

- For simply-supported beams and one-way slabs, at least $0.5A_{st}$ provided at mid-span should extend to the supports and have an effective anchorage of 12ϕ past the support centre. The remaining $0.5A_{st}$ should extend to within $0.08l$ of the support.
- For cantilever beams and one-way slabs, at least $0.5A_{st}$ used at the support should extend to the end of the cantilever. The remaining $0.5A_{st}$ should extend a distance $l/2$ or 45ϕ, whichever is the greater, from the support.
- For continuous beams of equal span, where the characteristic imposed load Q_k is not greater than the characteristic dead load G_k:
  (i) at least $0.2A_{st}$ over the supports should be made effectively continuous through the spans. Of the remainder, half should extend to a point not less than $0.25l$ from the support, and the other half to a point not less than $0.15l$ from the support, but no bar should stop at a point less than 45ϕ from the support;
  (ii) at least $0.3A_{st}$ at midspan should extend to the supports. The remainder should extend to points not less than $0.15l$ from the interior supports, and not less than $0.1l$ from the exterior supports;
  (iii) at a simply-supported end, the detailing rules for simply-supported beams should be applied.
- For continuous slabs of approximately equal span, where $Q_k \ngtr G_k$, 100 per cent of A_{st} should extend a distance of $0.1l$ or 45ϕ, whichever is the greater, and at least $0.5A_{st}$ should extend $0.3l$ into the span.

- The tension reinforcement at mid-span of a slab should extend to within $0.2l$ of the internal supports and within $0.1l$ of the external supports, and at least $0.5A_{st}$ should extend into the support.
- Where a slab is monolithic with its supporting beam at an end support, tension reinforcement equal to $0.5A_{st}$ at mid-span should be provided, and such bars should extend $0.1l$ or 45ϕ, whichever is the greater, into the span.
- The reinforcement spacing rules given in section 9.3 should be followed.

13.4 Sample detail drawings

Figs 13.2 to 13.9 on the following pages show sample detail drawings of:

- floor slabs;
- beams (floors and foundations or ground beams);
- staircases;
- bases (isolated and rafts); and
- a bending schedule.

These drawings do not necessarily cover the whole range of possible structural details but they have been included to serve as an introduction to the beginner in the art of detailing.

The interested student should study the drawings very closely. Although the details are by-and-large self-explanatory, the following points are worth noting.

(1) The style which is used to denote bars. For example:

- 13-R1002-300 (B) in Drg. 001 (Fig. 13.2) means 13 mild steel round bars, 10 mm in diameter, bar mark 2, at 300 mm centre-to-centre in the bottom of the slab;
- for a beam or column [see Drgs. 002 and 004 (Figs. 13.3 and 13.5)], R1201 means 12 mm mild steel round bars mark 1;
- for high-tensile bars, the symbol Y is used (instead of the symbol R for mild steel);
- for reinforcement placed at the top of a member the symbol T is used (instead of the symbol B for the bottom face).

(2) The dimensions are in mm. For example:

- 3600 in Drg. 001 (Fig. 13.2) means 3600 mm or 3.6 m;
- the description 225 × 450 on a beam in Drg. 001 (Fig. 13.2) implies the beam is 225 mm wide and 450 mm deep.

(3) The reinforcement within the span of a floor slab is placed at the bottom face (where tensile stresses are developed) and at the top face in the neighbourhood of the supports [see Drg. 001 (Fig. 13.2)].

(4) The reinforcement for isolated bases is placed at the bottom face [see Drg. 004 (Fig. 13.5)].

(5) The reinforcement in a solid raft is placed at the top face within the span, because the tension is now due to earth pressures [see Drg. 005 (Fig. 13.6)], and at the bottom at supports.

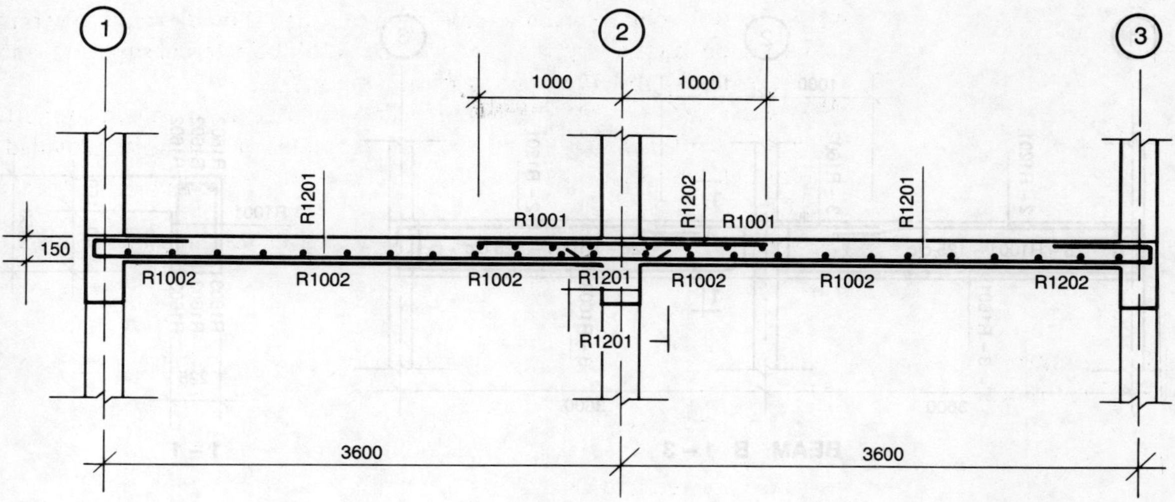

SECTION 1 – 1

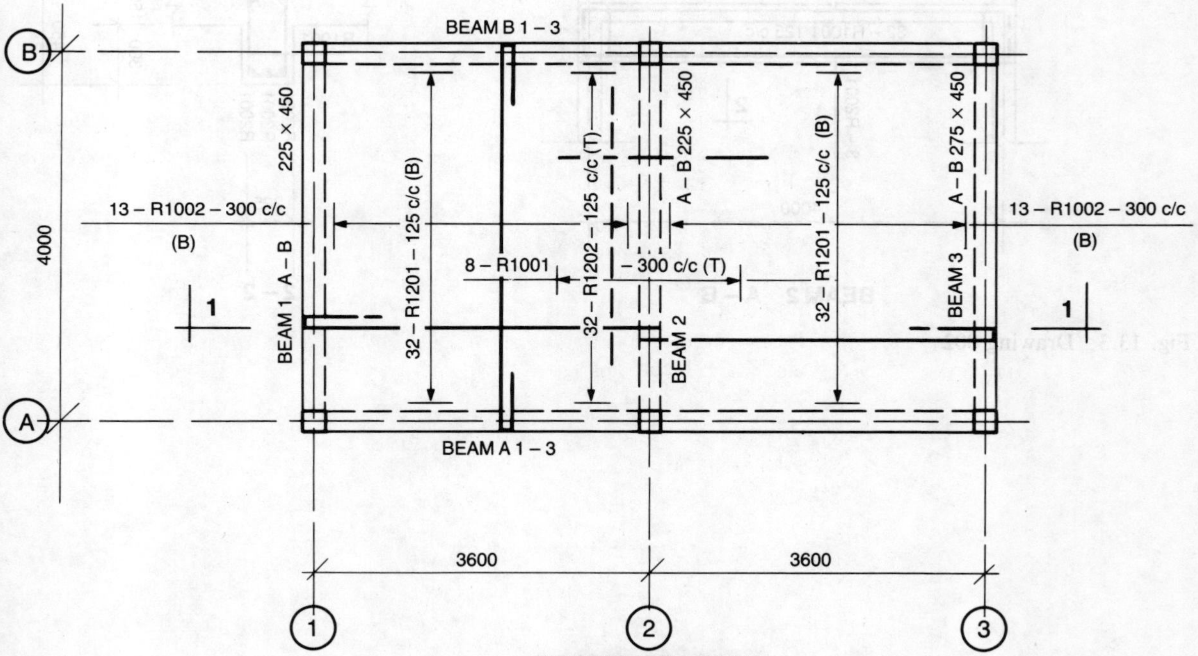

FIRST FLOOR SLAB

Fig. 13.2 Drawing 001

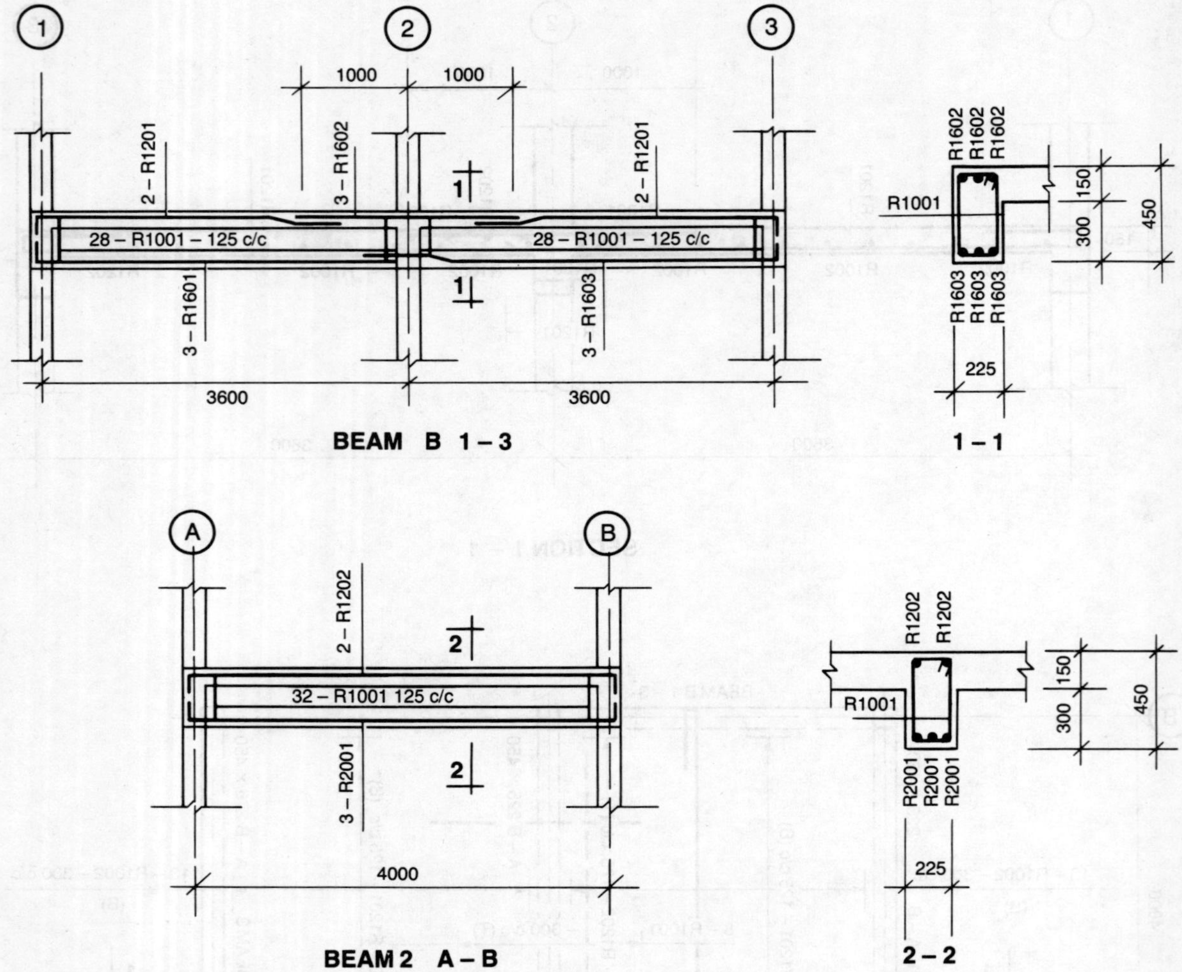

Fig. 13.3 Drawing 002

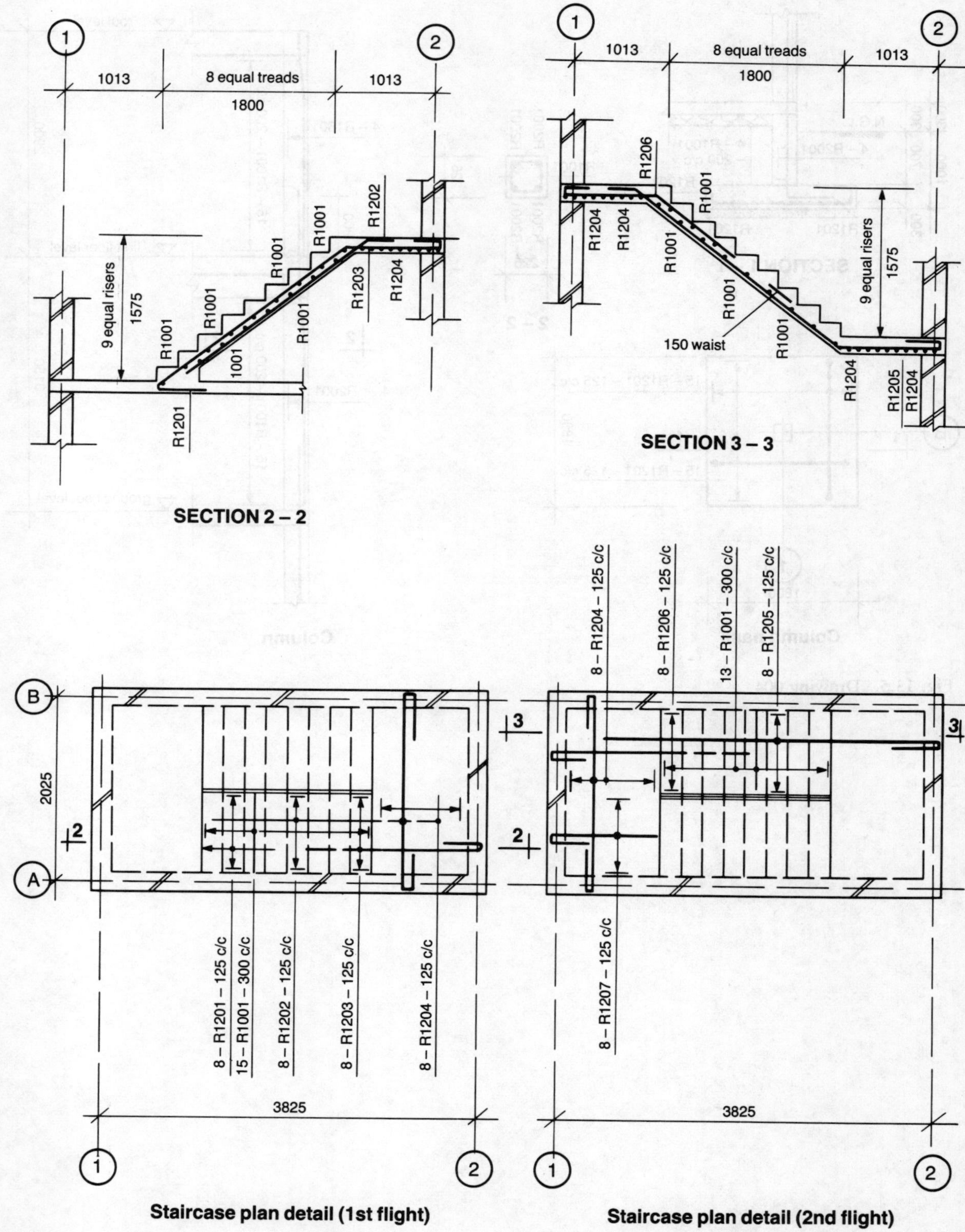

Fig. 13.4 Drawing 003

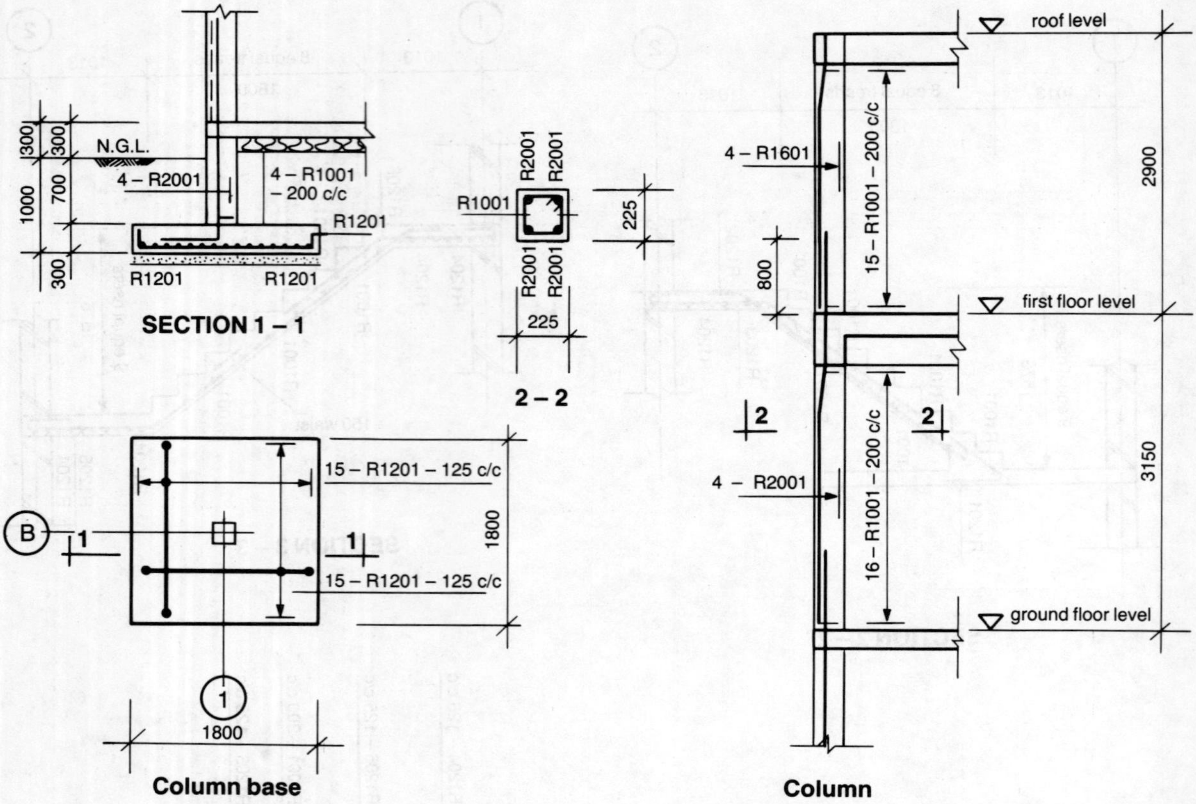

Fig. 13.5 Drawing 004

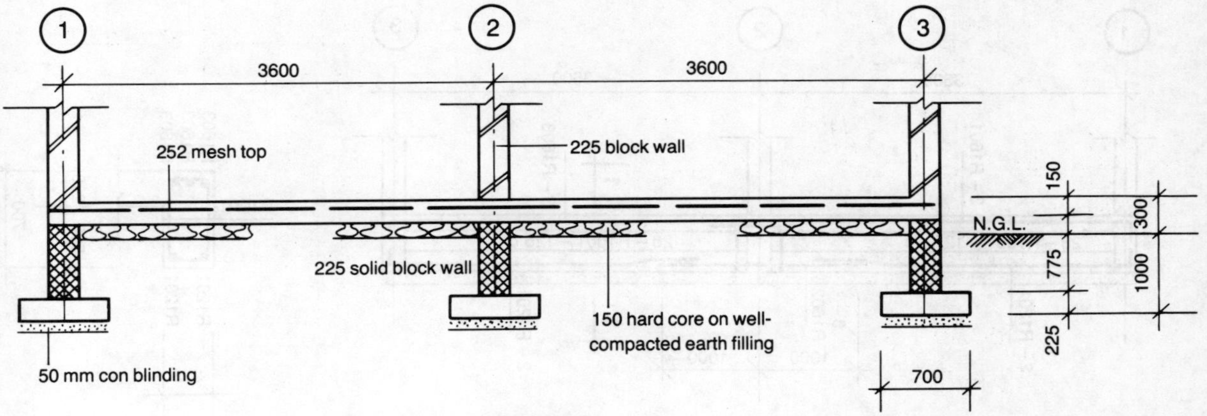

SECTION 3 – 3

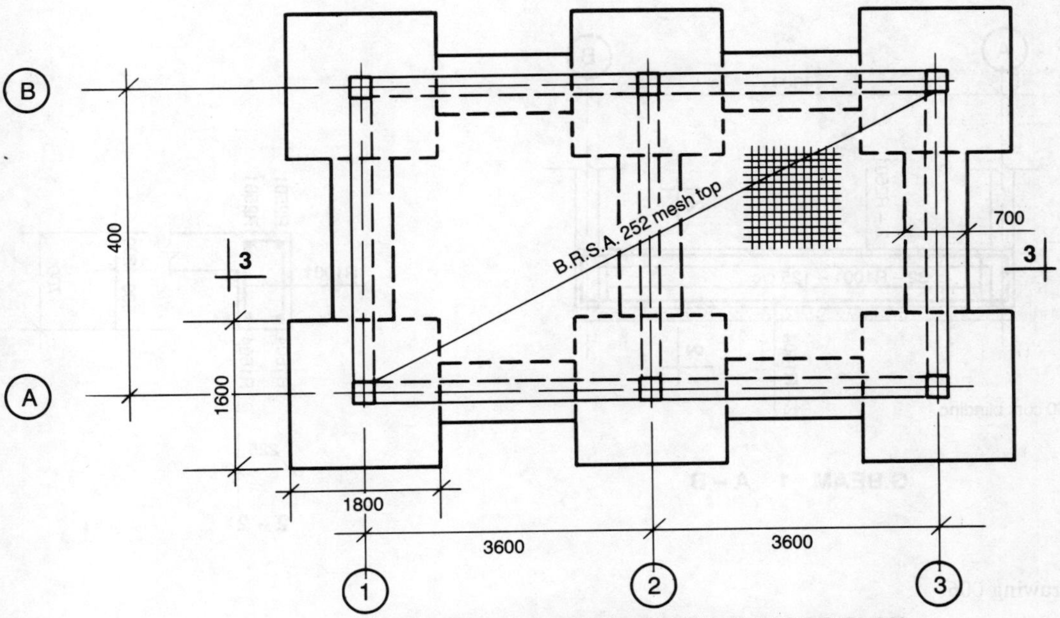

FOUNDATION AND GROUND FLOOR PLAN

Fig. 13.6 Drawing 005

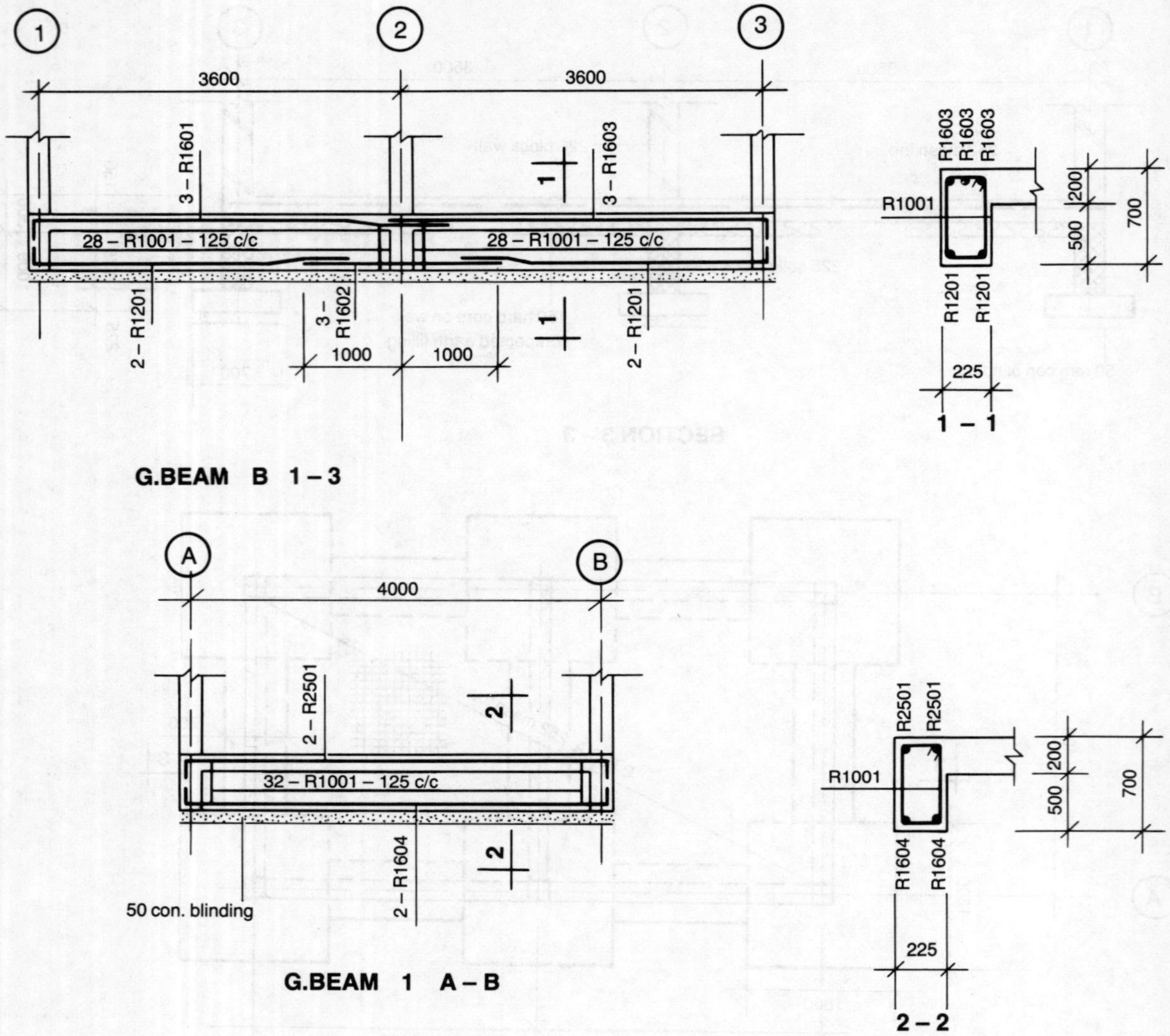

Fig. 13.7 Drawing 006

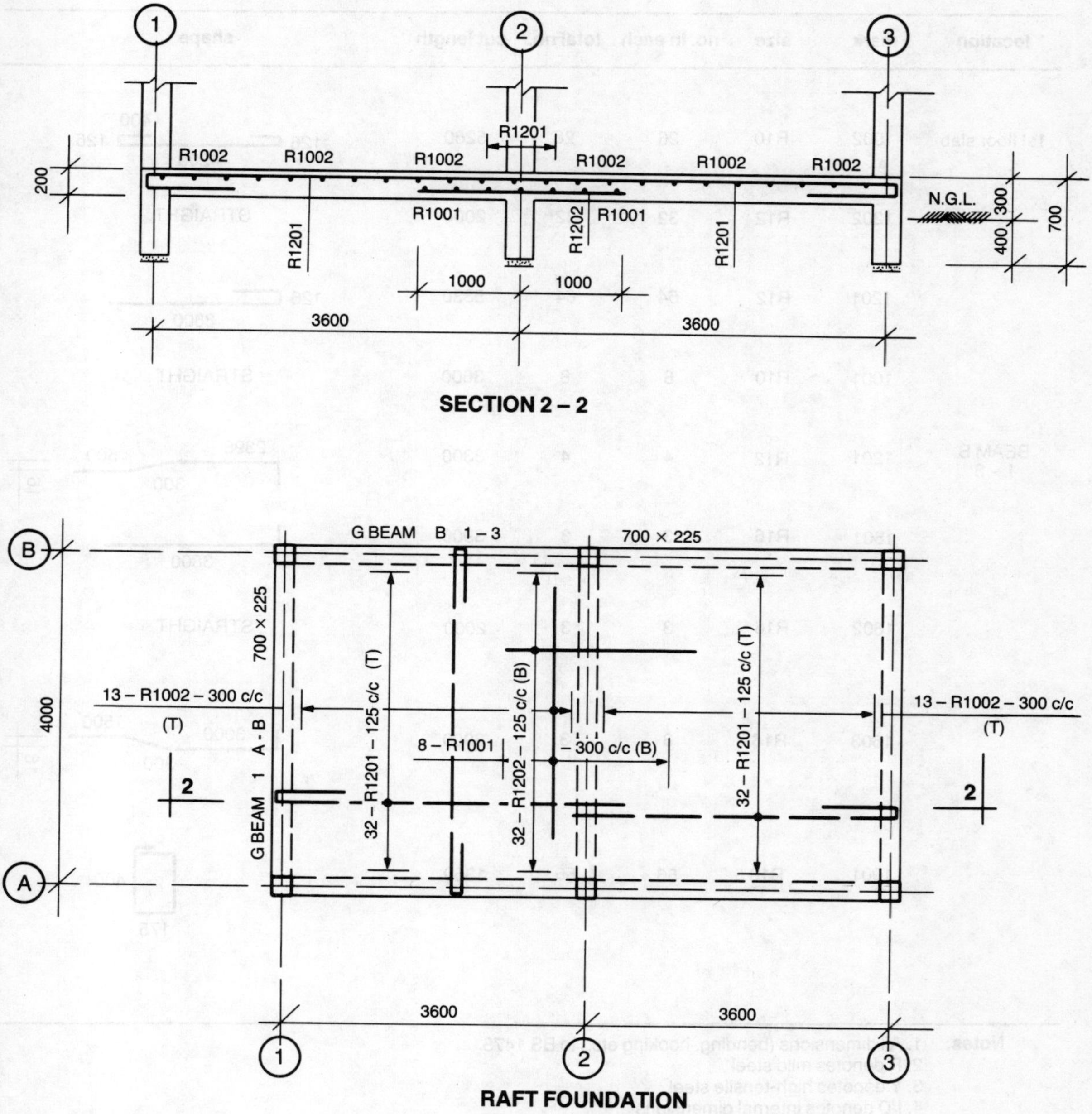

Fig. 13.8 Drawing 007

location	mark	size	no. in each	total no.	cut length	shape
1st floor slab	1002	R10	26	26	5260	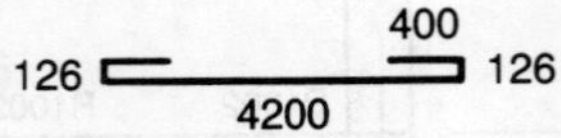
	1202	R12	32	32	2000	STRAIGHT
	1201	R12	64	64	5330	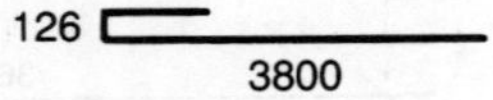
	1001	R10	8	8	3600	STRAIGHT
BEAM B 1 – 3	1201	R12	4	4	3300	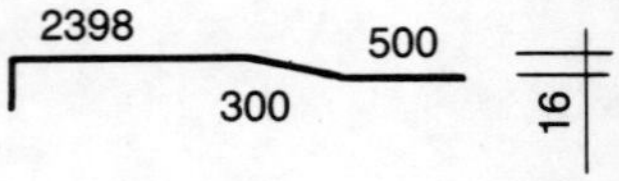
	1601	R16	3	3	3900	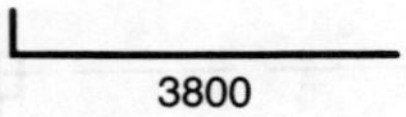
	1602	R16	3	3	2000	STRAIGHT
	1603	R16	3	3	3900	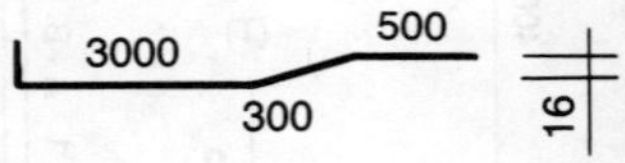
	1001	R10	56	56	1350	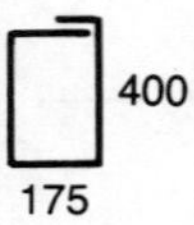

Notes:
1. All dimensions (bending, hooking etc.) to BS 1478
2. R denotes mild steel
3. Y denotes high-tensile steel
4. I/D denotes internal dimensions
5. O/A denotes overall dimensions

Fig. 13.9 Bar bending schedule

14 Computer-aided design

14.1 Introduction

The need to produce extensive designs of large and often complex structural systems, within a relatively short time, makes the use of computers in the modern engineering office imperative. A computer-aided design eliminates the tedious and often boring routine of performing repetitive calculations, leaving the engineer free to channel all mental effort to the philosophy of design. At the press of a button (literally) various designs of the same structural system can be carried out and the best (in terms of economy, strength, serviceability and even aesthetics) chosen.

When a comprehensive graphics package is available for a particular computer the failure mode of the structure for a large spectrum of loading combinations can be displayed on a video display unit (VDU) which is essentially a television screen. Sophisticated computer programs are also available for preparing structural detail drawings.

It is partly because of the need to prepare the student for computer-aided design methods that the author has deliberately avoided the use of design charts in this book. Obviously it is easier to base computer programs on the appropriate formulae rather than on charts derived from these equations.

The programs described and listed below have been developed by the author and his students over the years, and are mostly interactive in nature. The developers have tested and 'debugged' these programs to the extent of the type of sample design problems used. However, it is necessary to provide this note of caution:

WARNING: The author and his representatives bear no responsibility for any liability that may arise as a result of the use of any of the listed programs for analysis and/or design of any structural system.

At least one of the programs (P-FRAME) is most suited to being run on main-frame computers because of the rather large core memory required. Almost all the others can be run on micro-computer systems with a minimum core memory size of 64K bytes. Table 14.1 contains a general summary of the program function, capabilities, language and other relevant information regarding the programs.

Name of Program	Function	Language	Capabilities	Computer system
P-FRAME	Analysis of plane frames using the stiffness method	FORTRAN	Handles plane structural systems consisting of up to 40 elements. Dimension statements can be modified so as to handle more. Output includes direct thrust, shear and bending moments on elements. Several designs of the same structure with various loading combinations can be done in one run.	Main-frames (eg, IBM VM/370)
SLAB	Designs solid slabs using the modular-ratio, load-factor or limit state method depending on the user's choice.	BASIC	Can design several slabs at the same time. The slabs can be simply-supported, cantilever, constrained or continuous over supports.	Micro-computers (eg, Apple II plus, IBM PC, PC-XT and compatibles)
I-BASE	Designs isolated square bases supporting square columns using the modular-ratio method.	BASIC	The program provides a simple sketch of the base section after the design is completed. The sketch shows the number and location of reinforcing bars.	Micro-computers (eg, Apple II plus, IBM PC, PC-XT and compatibles)
RW-M1	Designs cantilever retaining walls using the modular-ratio and load-factor methods depending on the user's choice.	BASIC	The program performs automatic tapering of walls and chooses the most appropriate bar sizes based on the computed reinforcement amount.	Micro-computers (eg, Apple II plus, IBM PC, PC-XT and compatibles)

Table 14.1 General summary of program function, capabilities, language and other relevant information

14.2 Program P-FRAME

This program is developed specifically for the elastic analysis of plane frames using the stiffness method. P-FRAME can accommodate several load combinations on the same structure, so that various designs of the structure can be achieved within a single run of the program.

14.2.1 Input/output

The input data consists of the following:

- number of nodes, NOD;
- number of elements, NEL;
- number of load cases, LC;
- nodal data (coordinates, load/displacement condition and degrees of freedom);
- element data (cross-sectional area, area moment of inertia and modulus of elasticity);
- displacements, loads and applied moments for load cases.

The output data includes:

- nodal displacements (in the global x,y axes plus rotation about the z-axis) for the different load cases;
- thrust, shear, bending moments on elements for the different load cases.

14.2.2 Variable names and definitions

A matrix used to output element thrust, shear and bending moment
B matrix of element stiffness
C matrix used in assembling and partitioning, later used as KFF
D dummy vector used for global matrix
X nodal x-coordinates
Y nodal y-coordinates
R vector of "right-hand-side" forces in solution vector
ID1, ID2 indicators for load, displacement application
IDOF1, IDOF2, IDOF3 indicators for degrees of freedom
NO, NO2 ordering vectors for assembling and partitioning
A1 element cross-sectional area
A2 element moment of inertia
E element modulus of elasticity
H2, H, H1 vectors used for element stiffness
V vector of applied loads
U vector of displacements
N1 nodal number of end-1 of element
N2 nodal number of end-2 of element
G, G1 gamma matrices for rotation
S, T matrices used in computing element forces

Sign convention

Anti-clockwise moments are positive. Compression positive on end-1 of elements.

14.2.3 Flow chart

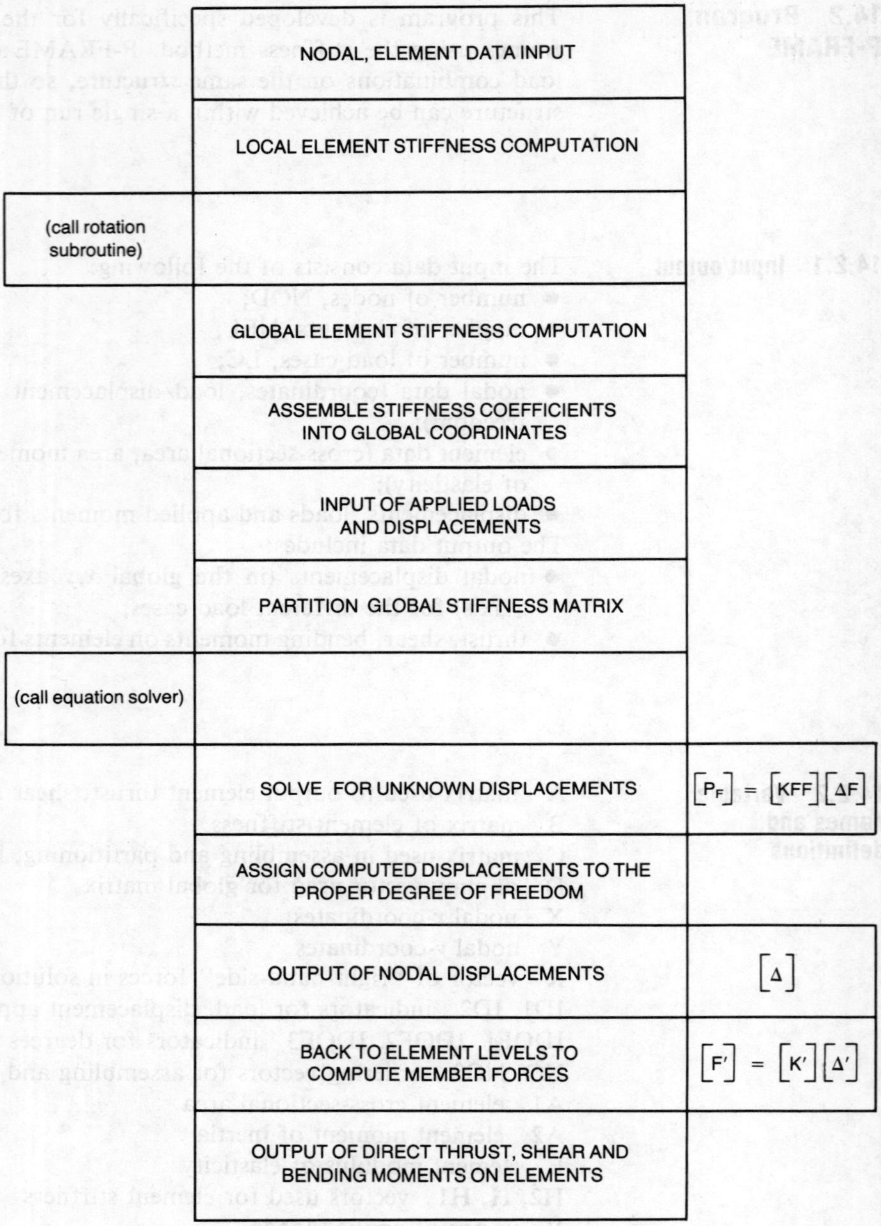

14.2.4 Specimen problem

Fig. 14.1 gives details of a specimen problem for solution using the P-FRAME program.

The specimen problem involves the analysis of a portal frame which carries a horizontal load P_2 and a bending moment M_2 at the joint 2 while transmitting another bending moment at the joint 3. Two load cases are considered. In the first case, the moments M_2 and M_3 are equal to zero, that is, the frame is carrying just the horizontal thrust. In the second case M_2 and M_3 are non-zero.

Fig. 14.1 Specimen problem for P-FRAME program

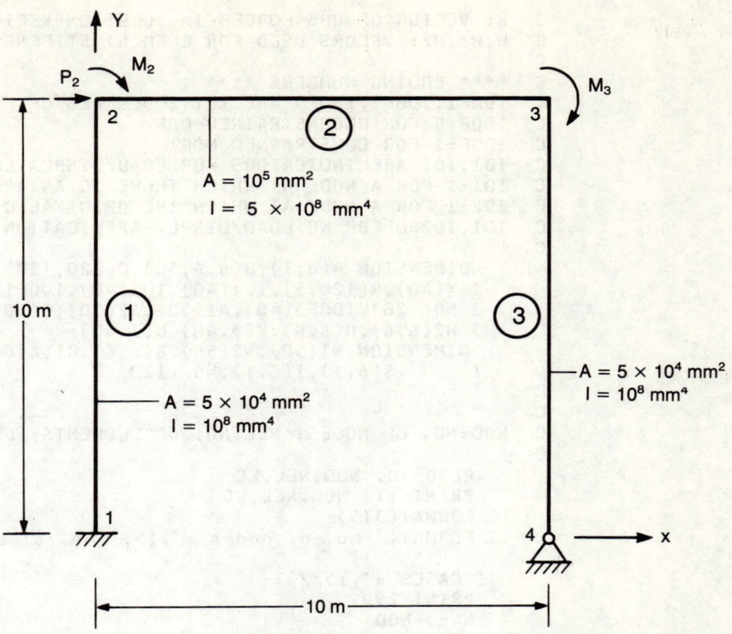

Load case	1	2
P_2	50 kN	50 kN
M_2	0	−5 kN m
M_3	0	−10 kN m

```
*WATFIV TIME=500
      IMPLICIT REAL *8(A-H,O-Z)
C
C THIS PROGRAM USES THE STIFFNESS METHOD FOR THE ANALYSIS OF PLANE
C FRAMES
C
C
C  **** UNITS ****
C  LENGTH MEASUREMENTS ARE IN METRES (M)
C  AREAS ARE IN SQUARE MILLIMETRES (MM2)
C  DIRECT AND SHEAR FORCES ARE IN KILO-NEWTONS (KN)
C  MOMENTS ARE IN KN-MM
C
C  **** ELEMENT PROPERTIES ****
C
C  A1: CROSS-SECTIONAL AREA
C  A2: MOMENT OF INERTIA
C   E: MODULUS OF ELASTICITY
C
C  **** VECTOR SYMBOLS ****
C
C  V: APPLIED FORCES
C  U: DISPLACEMENTS
C  B: MATRIX OF ELEMENT STIFFNESS
C  C: VECTOR USED IN ASSEMBLING AND PARTITIONING --
C     LATER USED AS THE KFF MATRIX
C  D: DUMMY VECTOR USED FOR GLOBAL MATRIX
C  G,G1: GAMMA MATRICES FOR ROTATION
```

```
C  R: VECTOR OF RHS FORCES IN SOLUTION VECTOR
C  H,H1,H2: VECORS USED FOR ELEMENT STIFFNESSES
C
C  **** CODING NUMBERS ****
C  IDOF1,IDOF2,IDOF3 ARE X,Y,Z DEGREES OF FREEDOM (DOF) RESPECTIVELY
C  IDOF=0 FOR UNCONSTRAINED DOF
C  IDOF=1 FOR CONSTRAINED DOF
C  ID1,ID2 ARE INDICATORS FOR LOAD/DISPLACEMENT APPLICATION
C  ID1=1 FOR A NODE AT WHICH THERE IS AN APPLIED LOAD
C  ID2=1 FOR A NODE AT WHICH THE ORIGINAL DISPL. IS PRESCRIBED
C  ID1,ID2=0 FOR NO LOAD/DISPL. APPLICATION
C
      DIMENSION A(6,1),B(6,6,50),C(120,120),D(120,120),X(40),
     1 Y(40),R(120,5),ID1(40),ID2(40),IDOF1(40),IDOF2(40),
     2 NO(120),IDOF3(40),A1(50),A2(50),E(50)
     3 H2(6,6),H(6,6),V(3,40),U(3,40)
      DIMENSION N1(50),N2(50),G(6,6),G1(6,6),H1(6,6),F1(3,3)
     1       ,S(6,1),T(6,1),NO2(120)
C
C
C  NOD=NO. OF NODES; NEL=NO. OF ELEMENTS; LC=NO. OF LOAD CASES
C
      READ 10, NOD,NEL,LC
      PRINT 11, NOD,NEL,LC
   10 FORMAT(3I5)
   11 FORMAT(` no. of nodes =`,i5,` no. of elements =`,i5,` no. of loa

     1 CASES =`,i5//)
      PRINT 2999
      NN=3*NOD
      DO 99 I=1,NN
      DO 99 J=1,NN
   99#(I,J)=0!
      LLC=1
C
C **** NODAL PROPERTIES INPUT ****
C
       DO 20 N=1,NOD
       READ 30,X(N),Y(N),ID1(N),ID2(N),IDOF1(N),IDOF2(N),IDOF3(N)
       DO 21 I=1,3
       V(I,N)=0!
       U(I,N)=0!
   21 CONTINUE
   20 CONTINUE
      PRINT 12
   12 FORMAT(`    nodal no    x-cood    y-cood`,8x,`  load:displ. cond.
     1    `,5x,degrees of freedom`)
      DO 13 N=1,NOD
   13 PRINT 14,N,X(N),Y(N),ID1(N),ID2(N),IDOF1(N),IDOF2(N),IDOF3(N)
   14 FORMAT(5X,I2,7X,D10.4,4X,D10.4,5X,I5,5X,I5,5X,I5,5X,I5,5X,I5/)
      PRINT 2999
   30 FORMAT(2D10.4,5I5)
   31 FORMAT(3D12.6)
C
 **** ELEMENT PROPERTIES INPUT ****

      DO 40 I=1,NEL
   40 READ 50,N1(I),N2(I),A1(I),A2(I),E(I)
      PRINT 15
   15 FORMAT(`    element no.  end-1  end-2  area    inertia
     1    AREA        MODULUS OF ELASTICITY`)
      DO 16 I=1,NEL
   16 PRINT 17,I,N1(I),N2(I),A1(I),A2(I),E(I)
   17 FORMAT(5X,I5,5X,I5,5X,I5,6X,D10.3,5X,D10.3,15X,D10.3/)
   50 FORMAT(2I5,3D10.3)
 2999 FORMAT(`-----------------------------------------------`//)
C
C **** COMPUTING STIFFNESS COEFFICIENTS IN LOCAL COODS. ****
C
      DO 70 I=1,NEL
      II=N1(I)
      JJ=N2(I)
      X1=X(II)
      X2=X(JJ)
      Y1=Y(II)
      Y2=Y(JJ)
```

```
      SL=1000!*DSQRT((X1-X2)**2+(Y1-Y2)**2)
      DO 80 M=1,6
      DO 80 MM=1,6
   80 B(M,MM,I)=0!
      B(1,1,I)=A1(I)*E(I)/SL
      B(1,4,I)=-B(1,1,I)
      B(2,2,I)=12!*E(I)*A2(I)/(SL**3)
      B(2,3,I)=6!*E(I)*A2(I)/(SL**2)
      B(2,5,I)=-B(2,2,I)
      B(2,6,I)=B(2,3,I)
      B(3,2,I)=B(2,3,I)
      B(3,3,I)=4!*E(I)*A2(I)/SL
      B(3,5,I)=-(B,2,3,I)
      B(3,6,I)=2!*E(I)*A2(I)/SL
      B(4,1,I)=B(1,4,I)
      B(4,4,I)=B(1,1.I)
      B(5,2,I)=B(2,5,I)
      B(5,3,I)=B(3,5,I)
      B(5,5,I)=B(2,2,I)
      B(5,6,I)=B(5,3,I)
      B(6,2,I)=B(2,6,I)
      B(6,3,I)=B(3,6,I)
      B(6,5,I)=B(5,6,I)
      B(6,6,I)=4!*E(I)*A2(I)/SL
      DO 84 K=1,6
      DO 84 KK=1,6
   84 H(K,KK)=B(K,KK,I)
C
C  **** ROTATING INTO THE GLOBAL COORDINATES ****
C
       CALL ROTA(X1,X2,Y1,Y2,G)
       CALL TRAN(G1,G,6,6)
       CALL MULT(H1,H,G,6,6,6)
       CALL MULT(H2,G1,H1,6,6,6)
C
C  **** ASSEMBLING STIFFNESS COEFFS. INTO GLOBAL COODS.
C
       IMIN=3*II-2
       KMIN=3*JJ-2
       DO 100 III=1,3
       DO 100 J=1,3
       D(IMIN+III-1,IMIN+J-1)=D(IMIN+III-1,IMIN+J-1)+H2(III,J)
       D(KMIN+III-1,IMIN+J-1)=D(KMIN+III-1,IMIN+J-1)+H2(III+3,J)
       D(IMIN+III-1,KMIN+J-1)=D(IMIN+III-1,KMIN+J-1)+H2(III,J+3)
       D(KMIN+III-1,KMIN+J-1)=D(KMIN+III-1,KMIN+J-1)+H2(III+3,J+3)
  100  CONTINUE
   70  CONTINUE
      DO 4444 LL=1,120
      DO 4444 LLL=1,120
 4444 C(LL,LLL)=0!
      IMX=3*NOD
C
C **** READING DISPL.,LOAD INPUT FOR LOAD CASES   ****
C
      LLC=1
   23 PRINT 24,LLC
      PRINT 5555
 5555 FORMAT(`  node no.`,5x,`x-load  (kn)`,10x,`y-load  (kn)`
     *      ,10X,Z-MOMENT  (KN-MM)`//)
   24 FORMAT(//`  load:displacement input for load case`,i2//)
      DO 22 N=1,NOD
      IF (ID1(N)O!EQ.O)GOTO 26
      READ 31,(V(IN,N),IN=1,3)
      PRINT 3304,N,V(1,N),V(2,N),V(3,N)
   26 IF (ID2(N)O!EQ.O)GOTO 22
      READ 31,(U(IN,N),IN=1,3)
      PRINT 3301
      PRINT 3304,N,U(1,N),U(2,N),U(3,N)
   22 CONTINUE
      PRINT 2999
C
C  **** PARTITIONING THE GLOBAL STIFFNESS MATRIX   ****
C
      III=1
      DO 1000 M=1,NOD
      J=3*M-2
```

```
      J1=IMX-J
      JMAX=J+2
      IF (IDOF1(M)O!EQ.1)GOTO 2200
      R(II1,LLC)=V(1,M)
      IF (ID2(M)O!EQ.1)R(II1,LLC)=V(1,M)-D(J,J)*U(1,M)
      NO(II1)=M
      IF (ID2(M)O!EQ.1)GOTO 2200
      DO 2300 JJ=1,IMX
 2300 C(II1,JJ)=D(J,JJ)
      NO2(II1)=J
      II1=II1+1
 2200 IF(IDOF2(M)O!EQ.1)GOTO 2400
      R(II1,LLC)=V(2,M)
      IF (ID2(M)O!EQ.1)R(II1,LLC)=V(2,M)-D(J+1,J+1)*U(2,M)
      NO(II1)=M
      IF(ID2(M)O!EQ.1)GOTO 2400
      DO 2500 JJ=1,IMX
 2500 C(II1,JJ)=D(J+1,JJ)
      NO2(II1)=J+1
      II1=II1+1
 2400 IF(IDOF3(M)O!EQ.1)GOTO 1000
      R(II1,LLC)=V(3,M)
      IF(ID2(M)O!EQ.1)R(II1,LLC)=V(3,M)-D(J+2,J+2)*U(3,M)
      NO(II1)=M
      J1M=J1-1
      DO 2700 JJ=1,IMX
 2700 C(II1,JJ)=D(JMAX,JJ)
      NO2(II1)=J+2
      II1=II1+1
 1000 CONTINUE
      IF(LC.EQ.1)GOTO 73
      IF(LLC.EQ.LC)GOTO 73
      LLC=LLC+1
      GOTO 23
   73 IIM=II1-1
      DO 72 I=1,IIM
      DO 72 J=1,IIM
      JJ=NO2(J)
   72#(I,J)=C(I,JJ)
      DO 1202 MM=1,IIM
      DO 1202 J=1,IIM
 1202 C(MM,J)=O!
      DO 1201 MM=1,IIM
      DO 1201 J=MM,IIM
 1201 C(MM,J-MM+1)=D(MM,J)
C
C  **** SOLVING FOR UNKNOWN DISPLACEMENTS  ****
C
      CALL SOLVE(C,R,IIM,IIM,!C)
C
C  **** ASSIGNING COMPUTED DISPL. TO RESPECTIVE DOFS
C
      LLC=0
 3001 LLC=LLC+1
      J=1
      PRINT 3002,LLC
 3002 FORMAT(//` output of results for load case`,i2//)
      DO 3000 I=1,NOD
      L=I
      IF(IDOF1(L)O!EQ.1)GOTO 3100
      U(1,L)=R(J,LLC)
      J=J+1
 3100 IF(IDOF2(L)O!EQ.1)GOTO 3200
      U(2,L)=R(J,LLC)
      J=J+1
      J=J+1
 3000 CONTINUE
C
C  **** OUTPUT OF NODAL DISPLACEMENTS  ****
C
      PRINT 3301
 3301 FORMAT(` node no.`,5x,`x-displ.  (mm)`,10x,y-displ. (mm)
     *  ,10X,`z-rotation (rad)`//)
      DO 3303 KKK=1,NOD
 3303 PRINT 3304,KKK,U(1,KKK),U(2,KKK),U(3,KKK)
 3304 FORMAT(I5,10X,D12.3,9X,D12.3,11X,D12.3//)
```

```
      PRINT 2999
C
C  **** GOING BACK TO ELEMENT LEVELS TO COMPUTE MEMBER FORCES ****
C
      DO 4000 I=1,NEL
        I1=N1(I)
        I2=N2(I)
        DO 4100 N=1,3
        T(N,1)=U(N,I1)
4100    T(N+3,1)=U(N,I2)
        X1=X(I1)
        X2=X(I2)
        Y1=Y(I1)
        Y2=Y(I2)
        CALL ROTA(X1,X2,Y1,Y2,G)
        CALL MULT(S,G,T,6,6,1)
        DO 4400 K=1,6
        DO 4400 KK=1,6
4400    H(K,KK)=B(K,KK,I)
        CALL MULT(A,H,S,6,6,1)
C
C  **** OUTPUT OF THRUST,SHEAR,BENDING MOMENT ON ELEMENT ****
C
        PRINT 6000,I,I1,I2
6000    FORMAT(//` element no.`,i2,10x,`node `,i3,10x,`node `,i3//)
        PRINT 6001,A(1,1),A(4,1)
        PRINT 6002,A(2,1),A(5,1)
        PRINT 6003,A(3,1),A(6,1)
6001    FORMAT(` direct thrust......`,d11.3,10x,d11.3,`    kn`)
6002    FORMAT(`             shear......`,d11.3,10x,d11.3,`    kn`)
6003    FORMAT(` bending moment.....`,d11.3,10x,d11.3,`   kn-mm`)
4000    CONTINUE
        PRINT 2999
        IF(LLC.LT.LC)GOTO 3001
        STOP
        END
C
C
C  *****  SUBROUTINES ------ SUBROUTINES ------ SUROUTINES *****
C
C
         SUBROUTINE ROTA(X1,X2,Y1,Y2,GAMA)
         IMPLICIT REAL*8(A-H,O-Z)
C
C  **** ROTA TRANSFORMS VECTORS FROM LOCAL TO GLOBAL COORDINATES ****
C
         DIMENSION GAMA(6,6)
         IF (Y1.EQ.Y2)GOTO 15
         IF (X1.EQ.X2)GOTO 10
         ALPH=DATAN((Y1-Y2 )/(X1-X2))
         GOTO 12
   15    IF(X1.GT.X2)GOTO 16
         ALPH=0!
         GOTO 12
   16    ALPH=3.141592654#
         GOTO 12
   10    IF(Y1.GT.Y2)GOTO 11
         ALPH=1.570796327#
         GOTO 12
   11    ALPH=4.712388981#
   12#0 13 I=1,6
         DO 13 J=1,6
   13    GAMA(I,J)=0!
         GAMA(1,1)=DCOS(ALPH)
         GAMA(1,2)=DSIN(ALPH)
         GAMA(2,1)=-DSIN(ALPH)
         GAMA(2,2)=DCOS(ALPH)
         GAMA(1,3)=0!
         GAMA(2,3)=0!
         GAMA(3,3)=1!
         GAMA(3,1)=0!
         GAMA(3,2)=0!
         DO 14 I=4,6
         DO 14 J=4,6
   14    GAMA(I,J)=GAMA(I-3,J-3)
         RETURN
         END
```

```
C
            SUBROUTINE MULT(A,B,C,M1,N1,N2)
            IMPLICIT REAL*8(A-H,O-Z)
C  MULT COMPUTES A=B*C
C  A,B,C ARE OF RESPECTIVE ORDERS M1XN2,M1XN1,AND N1XN2
C
            DIMENSION A(M1,N2),B(M1,N1),C(N1,N2)
            DO 80 I=1,N1
            DO 80 J=1,N2
            A(I,J)=0!
            DO 80 K=1,N1
         80 A(I,J)=A(I,J)+B(I,K)*C(K,J)
            RETURN
            END
C
            SUBROUTINE TRAN(A,B,M1,N1)
            IMPLICIT REAL*8(A-H,O-Z)
C
C  TRAN TRANSPOSES A GIVEN MATRIX B
C
            DIMENSION A(M1,N1),B(N1,M1)
            DO 10 I=1,N1
            DO 10 J=1,M1
         10 A(J,I)=B(I,J)
            RETURN
            END
C
            SUBROUTINE SOLVE(A,B,NN,MM,LC)
            IMPLICIT REAL*8(A-H,O-Z)
C
C  SOLVE IS A SYMMETRIC EQUATION SOLVER
C
C   A=MATRIX  STORED AS BAND
C   B=IPUT FORCE VECTOR, OUTPUT SOLUTION VECTOR
C   NN=NUMBER OF EQUATIONS
C   MM=BAND WIDTH;LC=WIDTH OF B
C
            DIMENSION A(120,120),B(120,5),C(100)
            N=0
        100 N=N+1
C           REDUCING N-TH EQUATION
C           DIVIDING RHS BY DIAGONAL ELEMENT
C
            DO 5 M=1,LC
          5 B(N,M)=B(N,M)/A(N,1)
C
C  CHECKING FOR LAST EQUATION
C
            IF (N-NN)151,300,151
C
C  DIVIDING N-TH EQUATION BY DIAGONAL ELEMENT
C
        151 IF (MM.LT.2)GOTO 100
        150#0 200 K=2,MM
              C(K)=A(N,K)
        200 A(N,K)=A(N,K)/A(N,1)
C
C  REDUCING THE REMAINING EQUATIONS
C
            DO 260 L=2,MM
            I=N+L-1
            IF (NN-I)260,240,240
        240 J=0
            DO 250 K=L,MM
            J=J+1
        250 A(I,J)=A(I,J)-C(L)*A(N,K)
           .DO 6 M=1,LC
          6 B(I,M)=B(I,M)-C(L)*B(N,M)
        260 CONTINUE
            GOTO 100
C
C BACK SUBSTITUTING
C
        300 N=N-1
            IF(N)350,500,350
        350 IF (MM.LT.2)GOTO 500
            DO 400 K=2,MM
```

```
      L=N+K-1
      IF(NN-L)400,370,370
370#0 7 M=1,LC
    7 B(N,M)=B(N,M)-A(N,K)*B(L,M)
400   CONTINUE
      GOTO 300
500   RETURN
      END
```

NO. OF NODES = 4   NO. OF ELEMENTS = 3   NO. OF LOAD CASES = 2

NODAL NO.	X-GOOD	Y-GOOD	LOAD:DISPL.		DEGREES OF FREEDOM		
1	0.0000D 00	0.0000D 00	0	0	1	1	1
2	0.0000D 00	0.1000D 02	1	0	0	0	0
3	0.1000D 02	0.1000D 02	1	0	0	0	0
4	0.1000D 02	0.0000D 00	0	0	1	1	0

ELEMENT NO.	END-1	END-2	AREA	INERTIA AREA	MODULUS OF ELASTICITY
1	1	2	0.500D 05	0.100D 09	0.200D 03
2	2	3	0.100D 06	0.500D 09	0.200D 03
3	3	4	0.500D 05	0.100D 09	0.200D 03

LOAD:DISPLACEMENT INPUT FOR LOAD CASE 1

NODE NO.	X-LOAD (KN)	Y-LOAD (KN)	Z-LOAD (KN-MM)
2	0.500D 02	0.000D 00	0.000D 00
3	0.000D 00	0.000D 00	

LOAD:DISPLACEMENT INPUT FOR LOAD CASE 2

NODE NO.	X-LOAD (KN)	Y-LOAD (KN)	Z-LOAD (KN-MM)
2	0.500D 02	0.000D 00	-0.500D 04
3	0.000D 00	0.000D 00	-0.100D 05

----THE FORCE VECTOR----

0.500D 02	0.500D 02
0.000D 00	0.000D 00
0.000D 00	-0.500D 04
0.000D 00	0.000D 00

```
                     0.000D 00       0.000D 00

                     0.000D 00      -0.100D 05

                     0.000D 00       0.000D 00
----------------------------------------------------

                 OUTPUT OF RESULTS FOR LOAD CASE 1

NODE NO.  X-DISPL. (MM)  Y-DISPL. (MM)  Z-ROTATION (RAD)

   1       0.000D 00      0.000D  00      0.000D  00

   2       0.185D 03      0.295D -01     -0.444D -02

   3       0.185D 03     -0.295D -01     -0.498D -03

   4       0.000D 00      0.000D  00     -0.276D -01
----------------------------------------------------


ELEMENT NO. 1         NODE 1          NODE 2
DIRECT THRUST        -0.295D 02       0.295D 02  KN
         SHEAR        0.392D 02      -0.392D 02  KN
 BENDING MOM.         0.205D 06       0.187D 06  KN-MM


ELEMENT NO. 2         NODE 2          NODE 3
DIRECT THRUST         0.108D 02      -0.108D 02  KN
         SHEAR       -0.295D 02       0.295D 02  KN
 BENDING MOM.        -0.187D 06      -0.108D 06  KN-MM


ELEMENT NO. 3         NODE 3          NODE 4
DIRECT THRUST         0.295D 02      -0.295D  02  KN
          SHEAR       0.108D 02      -0.108D  02  KN
 BENDING MOM.         0.108D 06       0.146D -10  KN-MM
----------------------------------------------------

                 OUTPUT OF RESULTS FOR LOAD CASE 2

NODE NO.  X-DISPL. (MM)  Y-DISPL. (MM)  Z-ROTATION (RAD)

   1       0.000D 00       0.000D  00     0.000D  00


   2       0.186D 03       0.310D -01    -0.446D -02

   3       0.186D 03      -0.310D -01    -0.100D -03

   4       0.000D 00       0.000D  00    -0.275D -01
----------------------------------------------------
```

```
ELEMENT NO. 1          NODE 1          NODE 2

DIRECT THRUST          -0.310D 02      0.310D  02  KN

        SHEAR          0.393D 02       -0.393D  02  KN

BENDING MOM.           0.205D 06       0.187D  06  KN-MM

ELEMENT NO. 2          NODE 2          NODE 3

DIRECT THRUST          0.107D 02       -0.107D  02  KN

        SHEAR          -0.310D 02      0.310D  02  KN

BENDING MOM.           -0.192D 06      -0.117D  06  KN-MM

ELEMENT NO. 3          NODE 3          NODE 4

DIRECT THRUST          0.310D 02       -0.310D  02  KN

        SHEAR          0.107D 02       -0.107D  02  KN

BENDING MOM.           0.107D 06       -0.146D -10  KN-MM
------------------------------------------------------

                      STATEMENTS EXECUTED = 8315
```

14.3 Program SLAB

SLAB is a program for the design of reinforced concrete rectangular solid slabs. The different types of slabs that can be designed include:

- simply-supported slabs;
- cantilevering slabs;
- slabs continuous over supports;
- slabs spanning in two directions.

14.3.1 Program capabilities

For each slab type, with the accompanying dimensions, support conditions, material properties and loading, the program:

- computes the moments and shear forces at critical sections;
- checks for shear, bond and span/depth ratios, and adjusts the slab thickness and reinforcement if necessary; and
- chooses suitable reinforcement for all parts of the slab.

SLAB is capable of designing several slabs within a single run. The program contains a variety of internal checks that ensure compliance with the requirements of the particular Code of Practice (selected by the user). If the requirements are not complied with the program takes the appropriate corrective measure and informs the user of the action.

SLAB is micro-computer based.

14.3.2 Program operation

The program begins with a series of questions that should be answered by the user. These include the following:

- the Code of Practice to be used. SLAB operates in accordance with the requirements of either CP 110 or CP 114 (which is equivalent to NCP 1). When the program is running based on CP 114 the user can choose either the modular-ratio method or the load-factor method.
- the characteristics of the slabs, such as the total number of slabs, slab types, the pertinent spans, and the end conditions.
- the material stresses and loads. The questions asked here depend on the Code selected.

14.3.3 Output

The output consists of:

- the designation (type) and number of each slab;
- the bar sizes and spacing;
- the placement of the reinforcement bars.

14.3.4 Flow chart

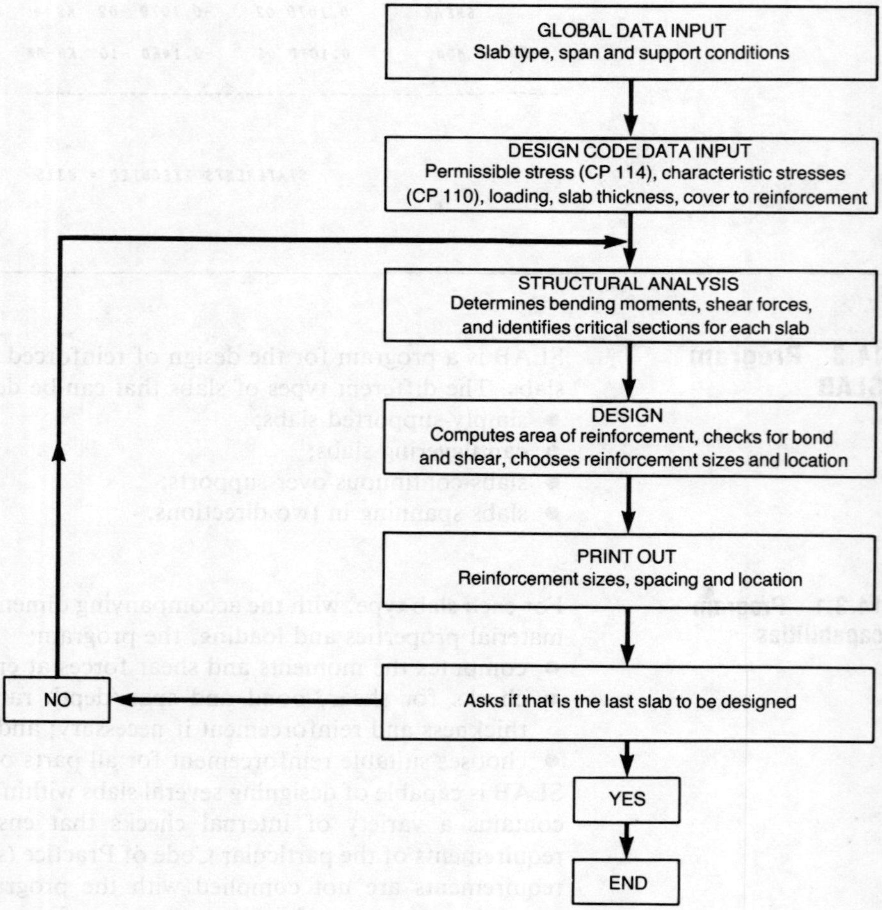

14.3.5 Specimen problem

The specimen problem for solution using the SLAB program is depicted in Fig. 14.2, the plan shown being one of a first-floor slab. The slab types may be divided as follows:

CANTILEVER SLAB	(1); VI
CONTINUOUS SLAB	(1); V
SIMPLY SUPPORTED SLAB	(2); I, IV
SLABS SPANNING IN TWO DIRECTIONS	(2); III, II

For entry into the computer, because there are six slabs:

Cantilever	1 is slab VI
Continuous slab	1 is slab V (2 SPANS)
Simply-supported slab	1 is slab IV
Simply-supported slab	2 is slab I
Two-way slab	1 is slab III
Two-way slab	2 is slab II

The spans of these slabs are entered into the computer, along with the support conditions.

The example given on the following pages has been carried out by all three methods, using mild steel and ordinary Portland 1:1:2 concrete. The method of entering the data can be seen by looking at the example, for which it has been assumed that:

dead load $= 6.6\,\text{kN/m}^2$
live load $= 2.0\,\text{kN/m}^2$
depth of slab $= 150$ mm
cover $= 20$ mm

Fig. 14.2 Specimen problem for SLAB program

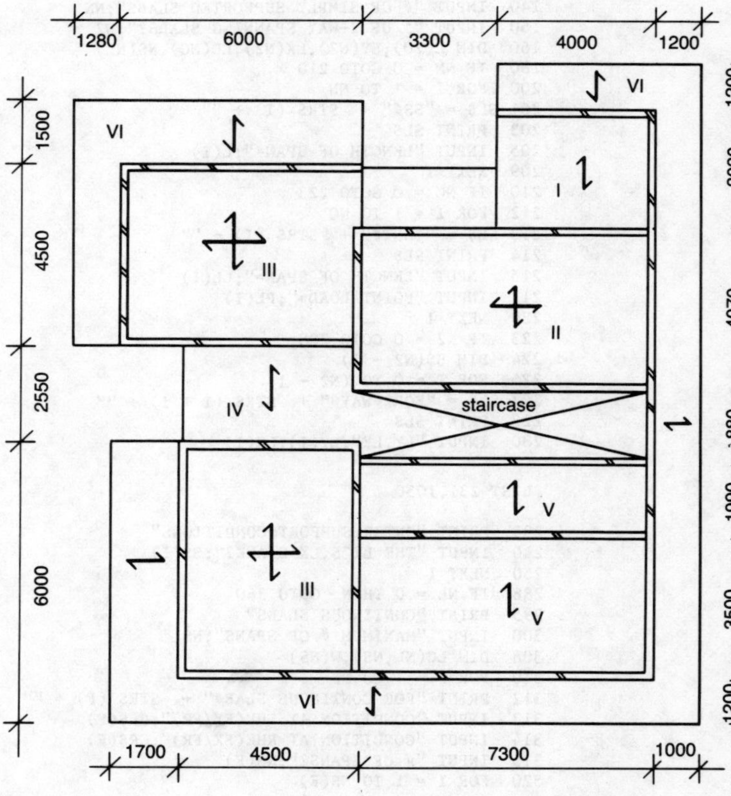

```
LOAD SLAB
]LIST 1,230

10  REM  THIS PROGRAM EXECUTES COMMERCIAL DESIGNS OF RECTANGULAR SLABS
11  REM  IT DESIGNS SLABS SPANNING IN TWO  DIRECTIONS AND SIMPLY SUPPORTED
      SLABS
12  REM  IT ALSO DESIGNS CANTILEVER SLABS AND SLABS CONTINOUS OVER MANY SU
      PPORTS
13  REM  IT CAN EXECUTE DESIGNS IN C.P.114 OR C.P.110
14  REM  IT IS OPERATED BY INITIALLY FEEDING IN ALL THE DATA REQUIRED FOR
      EVERY SLAB TO BE DESIGNED
15  REM  THE PROGRAM THEN PROCEEDS TO DESIGN ALL THE SLABS, MAKING ANY NEC
      ESSARY ADJUSTMENTS TO ENSURE COMPLIANCE WITH THE CODE OF PRACTICE
16  REM  THE PROGRAM SELECTS REINFORCEMENT SUITABLE FOR EACH SLAB
20  REM  ENTER MATERIAL STRESSES IN N/MM^2
21  REM  ENTER SLAB SPANS IN METRES
22  REM  ENTER LOADS IN KN/M^2
23  REM  ENTER SLAB DEPTHS IN MM
33 PI = 3.14159:JJ$ = "LOCAL BOND"
35  DIM ND(25)
40  DATA  5,5,5,5,5,8,5,5,5,5,5,12,12,12,12,12,12,12,12,12,5,5,7,5,7
42  FOR I = 0 TO 24
44 READ ND(I)
46  NEXT I
50  DIM Z(24,12)
53  FOR I = 0 TO 24
55  FOR J = 1 TO ND(I)
57  READ Z(I,J)
59  NEXT J
60  NEXT I
100  INPUT "CODE,114/110?";C$
105  DIM D1(4),D2(6),SZ(4),SP(6),DT$(1)
106  INPUT "ARE YOU USING A PRINTER(Y/N)?";PR$
107  IF PR$ = "Y" THEN  GOTO 110
108  PR# 0
110  INPUT "# OF SLABS?";N1
120  INPUT "# OF CANTILEVERS?";NC
130  INPUT "# OF CONTINOUS SLABS?";NL
140  INPUT "# OF SIMPLY SUPPORTED SLABS";NN
150  INPUT "# OF 2-WAY SPANNING SLABS?";N2
160  DIM L(20),LY(N2),LX(N2),LL(NC),NS(NL)
180  IF NN = 0 GOTO 210
200  FOR I = 1 TO NN
201 SL$ = "SS#" +  STR$ (I) + ""
203  PRINT SL$
205  INPUT "LENGTH OF SPAN=";L(I)
209  NEXT I
210  IF NC = 0 GOTO 223
212  FOR I = 1 TO NC
213 SL$ = "CANT#" +  STR$ (I) + ""
214  PRINT SL$
215  INPUT "LENGTH OF SPAN=";LL(I)
217  INPUT "POINT LOAD=";PL(I)
220  NEXT I
223  IF N2 = 0 GOTO 288
224  DIM S$(N2 - 1)
225  FOR I = 0 TO (N2 - 1)
227 SL$ = "FOR2-WAY#" +  STR$ (I + 1) + ""
228  PRINT SL$
230  INPUT "LY,LX";LY(I),LX(I)

LIST 231,1050

235  PRINT "ENTER SUPPORT CONDITIONS"
240  INPUT "THE LY´S,LX´S ARE?";S$(I)
250  NEXT I
288  IF NL = 0 THEN  GOTO 360
295  PRINT "CONTINOUS SLABS"
300  INPUT "MAXIMUM # OF SPANS";NS
308  DIM LC(NL,NS),W(NS)
310  FOR F = 1 TO NL
312  PRINT "FOR CONTINOUS SLAB#" +  STR$ (F) + ""
313  INPUT "CONDITION AT LHE(FX/FR)";CL$(F)
314  INPUT "CONDITION AT RHE(FX/FR)";CR$(F)
315  INPUT "# OF SPANS?";NS(F)
320  FOR I = 1 TO NS(F)
```

```
330 SL$ = "LENGTH OF SPAN#" + STR$ (I) + ""
332  PRINT SL$
335  INPUT LC(F,I)
340  NEXT I
350  NEXT F
355 F = 1
360  IF C$< > "110" THEN  GOTO 2000
380  GOTO 1000
692  PRINT  CHR$ (15)
800 IF PR$ = "Y" THEN  GOSUB 6200
802  FOR B = 0 TO NC
803  IF B = 0 GOTO 820
805  IF C$ = "114" THEN  GOSUB 2039
808  IF C$ = "114" GOTO 820
810  GOSUB 1141
820  NEXT B
830  FOR B = 0 TO NL
831 F = B
832  IF B = 0 GOTO 850
835  IF C$ = "114" THEN  GOTO 6100
838  IF C$ = "114" GOTO 850
840  GOSUB 1118
845  PRINT ""
850  NEXT B
860  FOR B = 0 TO NN
862  IF B = 0 GOTO 880
865  IF C$ = "114" THEN  GOSUB 2033
868  IF C$ = "114" GOTO 880
870  GOSUB 1152
875  PRINT ""
880  NEXT B
900  FOR B = 0 TO (N2 - 1)
903  IF N2 = 0 GOTO 950
905  IF C$ = "114" THEN  GOSUB 2037
908  IF C$ = "114" GOTO 920
910  GOSUB 1147
915  PRINT ""
920  NEXT B
930  PR# 0
950  END
1000  DIM X(4),Y(4),B(4)
1010 MD$ = "LS"
1015  INPUT "FY=";FY
1020  INPUT "FCU=";FU
1025 S = FY / FU
1030  INPUT "QK=";QK
1040  INPUT "GK=";GK
1043  INPUT "DEPTH=";DO
1050  INPUT "COVER=";DS:D1 = DO - DS:D9 = D1

 LIST 1051,1400

1055  IF TB$ = "P" THEN  GOTO 1080
1060  IF FU = 20 THEN  LET FB = 2.1
1065  IF FU = 25 THEN  LET FB = 2.5
1070  IF FU = 30 THEN  LET FB = 2.8
1075  IF FU = 40 THEN  LET FB = 3.4
1078  GOTO 1102
1080  IF FU = 20 THEN  LET FB = 1.7
1085  IF FU = 25 THEN  LET FB = 2.0
1090  IF FU = 30 THEN  LET FB = 2.2
1095  IF FU = 40 THEN  LET FB = 2.7
1102 W = 1.6 * QK + 1.4 * GK
1110  GOTO 800
1115  IF NL = 0 GOTO 1141
1118  FOR I = 1 TO NS(F) STEP 2
1120 W(I) = W: IF I = NS(F) THEN  GOTO 1122
1121 W(I + 1) = GK
1122  NEXT I
1123 T = 0
1125  GOTO 1138
1128  FOR I = 1 TO NS
1130 W(I) = GK
1132  NEXT I
1133 T = 1
1134 Y =  INT (NS / 2)
```

```
1136 W(Y) = W:W(Y + 1) = W
1138 SD = 26: GOTO 3300
1141 SD = 7: GOTO 3050
1147 SD = 26: GOTO 4600
1152 SD = 20: GOTO 3150
1159 N = 4
1160  IF DO = 150 THEN  LET FV = FV * 1.2
1165  IF DO = 175 THEN  LET FV = FV * 1.15
1170  IF DO = 175 THEN  LET FV = FV * 1.10
1220  IF MM > .15 * FU * 1000 * D1 ^ 2 THEN  GOTO 1250
1221  IF MM > .5 THEN  GOTO 1225
1222 SZ = 000:SP = 000:QB = 0
1223  GOTO 5400
1225 Q = MM * 100 / FU / D1 ^ 2
1230 R = 100 / S * (.4546 -  SQR (.2067 - 1.054 * Q))
1235 ST = 10 * R * D1
1240  GOTO 1300
1250  PRINT  TAB( 110)"DEPTH INSUFFICIENT"
1260 D1 = D1 + 25
1270  GOTO 1220
1300  DATA  .25,.5,1,2,3
1302  FOR I = 0 TO N
1310 X(I) = Z(6,I + 1)
1320  NEXT I
1330  DATA  2,1.98,1.44,1.13,1.01
1335  DATA  1.6,1.23,1,.84,.77
1340  DATA  1.55,1.2,.98,.83,.76
1345  DATA  1.41,1.11,.92,.78,.72
1350  IF FY = 250 THEN  LET RR = 1
1355  IF FY = 410 THEN  LET RR = 2
1360  IF FY = 425 THEN  LET RR = 3
1364  IF FY = 460 THEN  LET RR = 4
1366  FOR I = 0 TO N
1370 Y(I) = Z(RR,I + 1)
1380  NEXT I
1400 FC = 0

 LIST 1401,2000

1410  FOR K = 0 TO N
1420 B(K) = 1
1430  FOR I = 1 TO N
1435  IF I = K THEN  GOTO 1460
1440 B(K) = B(K) * (R - X(I)) / (X(K) - X(I))
1460  NEXT I
1470 FC = FC + Y(K) * B(K)
1475  NEXT K
1477  IF ZZ$ = "X" THEN  GOTO 1563
1479 F1 = FC
1480 SD = SD * FC
1485  IF SC < 50 GOTO 1500
1486 CC$ = "D"
1487  DATA  1.07,1.14,1.2,1.25,1.33,1.4,1.45,1.5
1495  GOTO 1407
1498 CC$ = ""
1500  IF L * 1000 / D1 < SD GOTO 1510
1503 PK = KP
1504 D1 = D1 + 25:SD = SD / FC
1505  GOTO 1220
1510 PK = KP
1511  GOTO 5000
1515  DATA  .25,.5,1,2,3
1516 RR = 6
1517  FOR I = 0 TO 4
1525 X(I) = Z(6,I + 1)
1527  NEXT I
1530  DATA  .35,.45,.6,.8,.85
1535  DATA  .35,.5,.65,.85,.9
1538  DATA  .35,.55,.7,.9,.95
1540  DATA  .35,.55,.75,.95,1
1542  IF FU = 20 THEN  LET RR = 7
1545  IF FU = 25 THEN  LET RR = 8
1546  IF FU = 30 THEN  LET RR = 9
1547  IF FU = 40 THEN  LET RR = 10
1548  FOR I = 0 TO 4
1550 Y(I) = Z(RR,I + 1)
```

```
1551  NEXT I
1558 ZZ$ = "X";N = 4
1561  GOTO 1400
1563 ZZ$ = "";N = 4;F2 = FC
1565 FV = FC
1568 QB = V * SP / PI / D1 / SZ
1570  IF QB > FB THEN 6010
1582  GOTO 5400
1590 SV = V / DO
1592  IF DO = 150 THEN  LET FV = FV * 1.2
1595  IF DO = 175 THEN  LET FV = FV * 1.15
1598  IF DO = 200 THEN  LET FV = FV * 1.1
1600  IF SV < = FV THEN  GOTO 1620
1605  PRINT "FAILED IN SHEAR"
1610  GOTO 1156
1620  PRINT "SHEAR O.K."
1650  IF CR$ = "COMP RE" THEN  LET S1 = ST;ST = SC
1700  GOTO 1015
2000  IF MM < > 0 THEN  GOTO 800

 LIST 2001,3000

2005  INPUT "ELASTIC/LOAD FACTOR(EL/LF)?";MD$
2010  INPUT "PCB=";PC
2013  INPUT "PST=";PS
2014  INPUT "DEPTH=";DO
2015  INPUT "COVER=";DS:D1 = DO - DS:D9 = D1
2016 H = PS - 15: IF H > 175 THEN  LET H = 175
2017  INPUT "LOAD=";W
2020  IF PC < > 10 GOTO 2035
2021 C = 1.5
2025  IF PC < > 8.5 GOTO 2030
2026 C = 1.4
2030  IF PC < > 7 GOTO 2032
2031 C = 1.25
2032  GOTO 800
2033 D = 1:SD = 30: GOTO 2040
2034  IF ST$ < > "
2035 D = 2:SD = 40: GOTO 2040
2037 D = 3:SD = 40: GOTO 2040
2039 D = 4:SD = 12: GOTO 2040
2040  IF PS > 200 THEN  LET SD = SD * .9
2044  ON D GOTO 3150,3300,4600,3050
2050  IF MM > .5 THEN  GOTO 2055
2052 SZ = 0:SP = 0:QB = 0
2053  GOTO 5400
2055  IF MD$ = "LF" GOTO 2150
2060 PK = KP
2070 M = 15:N = 1 / (1 + PS / M / PC):A = 1 - N / 3:K = .5 * PC * N * A
2080 MR = K * D1 ^ 2 / 10 ^ 3
2085  IF MR < MM THEN  GOTO 2097
2090 ST = MM * 10 ^ 6 / PS / A / D1
2095  GOTO 2300
2097 PK = KP
2100 D1 = D1 + 25
2130  GOTO 2080
2150 K = MM * 10 ^ 6 / (1000 * D1 ^ 2)
2155  IF K > PC / 4 THEN  GOTO 2200
2165 RO = 2 * PC * (1 -  SQR (1 - 3 * K / PC)) / 3 / PS
2170 N = 3 * PS * RO / 2 / PC
2180 A = 1 - N / 2
2190 ST = 1000 * RO * D1
2195  GOTO 2300
2200 D1 = D1 + 25
2230  GOTO 2150
2300  GOTO 5000
2305 QB = V * SP / PI / SZ / A / D1
2310  IF QB > C THEN 6010
2330  GOTO 5400

 LIST 3001,3500

3050 SS$ = "FOR CANTILEVER #" +  STR$ (B) + "        "
3058 PL = PL(B)
3060 L = LL(B)
```

```
3065 MM = W * L ^ 2 / 2
3068 MM = MM + L * PL
3070 B$ = "ALONG TOP"
3075 V = W * L + PL
3078  IF C$ = "110" THEN  GOTO 3100
3080  IF SD > L * 1000 / D1 GOTO 3095
3085 KP = PK
3086 D1 = D1 + 25
3088  GOTO 3080
3095 PK = KP
3100  IF C$ = "110" GOTO 1159
3110  GOTO 2050
3150 SS$ = "FOR SS#" + STR$ (B) + "                "
3155 L = L(B)
3165 MM = W * L ^ 2 / 8
3170 B$ = "ALONG BOTTOM"
3180 V = W * L / 2
3181  IF C$ = "110" GOTO 3190
3182  IF SD > L * 1000 / D1 GOTO 3188
3183 PK = KP
3185 D1 = D1 + 25
3187  GOTO 3182
3188 KP = PK
3190  IF C$ = "110" GOTO 1159
3200  GOTO 2050
3300  PRINT "   "
3301  IF F > 1 GOTO 3487
3302  IF T > 0 THEN  GOTO 3487
3303 NS = NS + 1
3304  DIM AM(NS),HM(NS),SS(NS),EC(NS),V(NS),MR(NS),ML(NS),DR(NS),DL(NS),SL
(NS)
3305  DIM WI(NS)
3306  DIM SR(NS),QR(NS),QL(NS),CR(NS),CL(NS),OM(NS),TR(NS),TL(NS)
3309 NS = NS - 1
3400 SS$ = "FOR CONTINOUS SLAB " +  STR$ (F) + "     "
3420 NS = NS(F)
3450  FOR I = 1 TO NS
3452 L(I) = LC(F,I)
3470  NEXT I
3483 CL$ = CL$(F):CR$ = CR$(F)
3487  IF CR$ = "FR" THEN  LET CL(NS) = 0
3488  IF CL$ = "FR" THEN  LET CR(0) = 0
3495  FOR I = 1 TO NS
3496  IF I = NS GOTO 3520
3500 DR(I) = (1 / L(I + 1)) / (1 / L(I + 1) + 1 / L(I))

 LIST 3501,4000

3510 DL(I) = 1 - DR(I)
3520  NEXT I
3522  IF CR$ = "FR" THEN  LET DR9R(NS - 1) = (.75 / L(NS)) / (.75 / L(NS) +
1 / L(NS - 1))
3525  IF CL$ = "FR" THEN  LET DL(1) = (.75 / L(1)) / (1 / L(2) + .75 / L(1
))
3530 DL(0) = 0:DL(NS) = 0:DR(NS) = 0
3540  FOR I = 1 TO (NS - 1)
3560 MR(I) =  - W(I + 1) * L(I + 1) ^ 2 / 12
3570 ML(I) = W(L) + L(I) ^ 2 / 12
3580 SR(I) = MR(I):SL(I) = ML(I)
3595  NEXT I
3610  IF CL$ = "FX" GOTO 3640
3620 MR(0) = 0:ML(0) = 0:ML(1) = W(1) * L(1) ^ 2 / 8
3625 SL(1) = ML(1)
3630  GOTO 3660
3640 MR(0) =  - W(1) * L(1) ^ 2 / 12
3643 ML(1) =  - MR(0)
3660  IF CR$ = "FX" GOTO 3700
3670 MR(NS - 1) =  - W(NS) * L(NS) ^ 2 / 8
3680  GOTO 3725
3700 MR(NS - 1) =  - W(NS) * L(NS) ^ 2 / 12
3710 ML(NS) =  - MR(NS - 1)
3725 SL(1) = ML(1):SR(0) = MR(0)
3730 SR(NS - 1) = MR(NS - 1):SL(NS) = ML(NS)
3740 J = 1
3750  FOR I = 0 TO NS
3760 TR(I) = MR(I):TL(I) = ML(I)
```

```
3770 QR(I) = MR(I):QL(I) = ML(I)
3800  NEXT I
3810  FOR I = 0 TO NS
3820 OM(I) = TR(I) + TL(I)
3830 QR(I) =  - OM(I) * DR(I)
3840 QL(I) =  - OM(I) * DL(I)
3845  NEXT I
3850  FOR I = 0 TO NS
3857 CR(I) = QL(I + 1) / 2
3858  IF I = 0 GOTO 3865
3860 CL(I) = QR(I - 1) / 2
3865  IF CR$ = "FR" THEN  LET CL(NS) = 0
3870  IF CL$ = "FR" THEN  LET CR(0) = 0
3880 SL(I) = SL(I) + QL(I) + CL(I)
3890 SR(I) = SR(I) + QL(I) + CL(I)
3893 TR(I) = CR(I):TL(I) = CL(I)
3900  NEXT I
3910 J = J + 1
3920  IF J < 6 THEN  GOTO 3810
3930 SL(0) =  - SR(0)
3940  FOR Z = 0 TO NS
3950 LM$ = "X"
3960 M$ = "M(STR$(Z))"
3975  IF T = 1 THEN  GOTO 3990
3980 X(Z) = SL(Z)
3985 WI(Z) = W(Z)
3990  NEXT Z
3991  IF T = 0 THEN  GOTO 1128
3992  FOR Z = 0 TO NS
3993  IF WI(Z) > W(Z) THEN  LET W(Z) = WI(Z)
3994  IF X(Z) > SL(Z) THEN  LET SL(Z) = X(Z)

 LIST 4001,4620

4055 SS(0) = W(1) * L(1) / 2
4060 EC(0) = (SL(0) - SL(1)) / L(1)
4070 SS(NS) = W(NS) * L(NS) / 2
4080 EC(NS) = (SL(NS) - SL(NS - 1)) / L(NS)
4100  FOR I = 1 TO NS
4105  IF I = 0 GOTO 4130
4107  IF I = NS GOTO 4130
4110 SS(I) = W(I) * L(I) / 2 + W(I + 1) * L(I + 1) / 2
4120 EC(I) = (SL(I) - SL(I - 1)) / L(I) + (SL(I) - SL(I + 1)) / L(I + 1)
4130 V(I) = SS(I) + EC(I)
4150 V = V(Z)
4170  NEXT I
4180  FOR Z = 0 TO NS
4182 B$ = "OVER SUPPORT #" +  STR$ (Z) + " "
4185 MM = SL(Z):V = V(Z)
4190  IF MD$ = "LS" GOTO 4196
4192  GOTO 2050
4196  GOTO 1159
4200  NEXT Z
4300 LM$ = "Z"
4305  FOR Z = 1 TO NS
4310 AM(Z) = (SL(Z) + SL(Z - 1)) / 2
4330 HM(Z) = W(Z) * L(Z) ^ 2 / 8 - AM(Z)
4333 B$ = "FOR SPAN #" +  STR$ (Z) + " "
4334 V = 0
4336 MM = HM(Z)
4340  IF MD$ = "LS" GOTO 4346
4342  GOTO 2050
4346  GOTO 1159
4350  NEXT Z
4353 LM$ = ""
4480  IF F = NL THEN  RETURN
4500  GOTO 3301
4600 SS$ = "2-WAY SLAB #" +  STR$ (B + 1) + "                "
4605  PRINT SS$
4610 LY = LY(B):LX = LX(B)
4611  IF C$ = "110" GOTO 4620
4612  IF SD > LX * 1000 / D1 GOTO 4618
4613  PRINT "SDR TOO LARGE"
4615 D1 = D1 + 25
4617  GOTO 4612
4618  PRINT "SDR IS O.K."
4620 N = 4
```

```
LIST 4621,4920

4622  IF C$ = "110" GOTO 4626
4623  IF TA > 0 GOTO 4632
4625 KP = PK
4626  IF TA > 0 THEN  GOTO 4632
4628  DIM T(N),U(N)
4632 R = LY / LX
4650 S$ = S$(B)
4668 RR = 0
4670  IF S$ = "CCCC" THEN  LET RR = 1
4672  IF S$ = "CCCD" THEN  LET RR = 2
4674  IF S$ = "CCDD" THEN  LET RR = 5
4676  IF S$ = "CDCC" THEN  LET RR = 3
4678  IF S$ = "CDCD" THEN  LET RR = 4
4680  IF S$ = "CDDD" THEN  LET RR = 7
4682  IF S$ = "DDCC" THEN  LET RR = 6
4684  IF S$ = "DDCD" THEN  LET RR = 8
4686  IF S$ = "DDDD" THEN  LET RR = 9
4703 T = 0:U = 0
4706  DATA  .032,.024,.043,.032,.051,.039,.060,.045,.065,.049,.032,.034
4708  DATA  .037,.028,.048,.036,.055,.041,.064,.048,.068,.052,.037,.028
4710  DATA   .037,.028,.052,.039,.063,.047,.077,.059,.085,.065,.037,.028
4712  DATA  .047,.035,.060,.045,.071,.053,.084,.063,.091,.069,.042,.035
4714  DATA  .045,.035,.052,.040,.059,.044,.065,.049,.069,.052,0,.035
4716  DATA  0,.035,.00,.051,0,.063,0,.080,0,.088,.045,.035
4717  DATA  .057,.043,.071,.053,.080,.060,.091,.069,.097,.073,0,.043
4718  DATA  0,.043,0,.059,0,.071,0,.087,0,.096,.057,.043
4720  DATA  .062,.062,.084,.059,.099,.051,.113,.037,.118,.029,0,.00
4722  FOR I = 0 TO N
4725 T(I) = Z(RR + 10,2 * I + 1):U(I) = Z(RR + 10,2 * I + 2)
4728  NEXT I
4729 L = Z(RR + 10,11):M = Z(RR + 10,12)
4730  DATA  1.0,1.2,1.4,1.75,2
4735  FOR I = 0 TO N
4742 X(I) = Z(20,I)
4750  NEXT I
4755  FOR K = 0 TO N
4756 B(K) = 1
4760  FOR I = 0 TO N
4763  IF I = K THEN  GOTO 4780
4770 B(K) = B(K) * (R - X(I)) / (X(K) - X(I))
4780  NEXT I
4790 T = T + T(K) * B(K)
4795 U = U + U(K) * B(K)
4800  NEXT K
4830 Q = W * LX ^ 2
4832  IF RR = 9 GOTO 4880
4835 M1 = T * Q:M2 = U * Q:M3 = L * Q:M4 = M * Q
4850 V = W * (1 - 1 / 2 / R) / 2 * LX
4875  GOTO 4900
4880 M2 = T * Q:M4 = U * Q:M1 = 0:M3 = 0
4900 TA = 1
4905  ON TA GOTO 4910,4915,4920,4925
4910 MM = M1:B$ = "OVER X SUPPORT"
4911 TA = TA + 1
4912  IF MD$ = "LS" GOTO 1159
4913  GOTO 2050
4915 MM = M2:B$ = "AT MID-SPAN (X)"
4916 TA = TA + 1
4917  IF MD$ = "LS" GOTO 1159
4918  GOTO 2050
4920 MM = M3:B$ = "OVER Y SUPPORT"
```

```
LIST 4921,6000

4921 TA = TA + 1
4922  IF MD$ = "LS" GOTO 1159
4923  GOTO 2050
4925 MM = M4
4926 TA = TA + 1:B$ = "AT MID-SPAN (Y)"
4927  IF MD$ = "LS" GOTO 1159
5000  DATA  10,10,6,3,1
5020  FOR I = 0 TO 4
5050 D1(I) = Z(21,I + 1)
5060  NEXT I
5070  DATA  1,2,6,4,10,8,10
5080  FOR I = 0 TO 6
5100 D2(I) = Z(22,I + 1)
5110  NEXT I
5115  DATA  10,12,16,20,25
5120  FOR I = 0 TO 4
5140 SZ(I) = Z(23,I + 1)
5150  NEXT I
5170 HP = 0
5200  DATA  300,250,200,175,150,125,100
5210  FOR I = 0 TO 6
5230 SP(I) = Z(24,I + 1)
5240  NEXT I
5250  IF SZ = 10 THEN  GOTO 5252
5251  IF BB$ = "F" THEN  LET D1(Y) = 0
5252 HP = 0
5255  FOR U = 0 TO 4
5258 M =  - 1
5270 SZ = SZ(U)
5280 ZZ = 250 * PI * SZ ^ 2 / ST
5290  IF M > 5 THEN  GOTO 5350
5300 M = M + 1
5310 SP = SP(M):DF = ZZ - SP
5315  IF DF < 0 GOTO 5290
5330  IF D1(U) * D2(M) < = HP GOTO 5370
5340  IF D1(U) * D2(M) > HP THEN  LET HP = D1(U) * D2(M)
5350 Y = U:X = M
5370  NEXT U
5380 SZ = SZ(Y):SP = SP(X)
5382  IF C$ = "110" GOTO 1568
5385  GOTO 2305
5400  IF PR$ = "Y" THEN  GOTO 6900
5402 DT$(0) = "USE" +  STR$ (SZ) + "MM BARS AT" +  STR$ (SP) + "MM SPACING
     "
5405  PRINT SS$
5408  PRINT "DEPTH";D1 + DS;"MM"
5410  PRINT DT$(0)
5420  PRINT B$
5430  PRINT
5435 D1 = D9
5490  IF TA = 5 THEN 6000
5495  IF TA <  > 0 THEN 4905
5500  IF LM$ = "X" THEN  GOTO 4200
5505  IF LM$ = "Z" AND (Z = NS) THEN  RETURN
5510  IF LM$ = "Z" THEN  GOTO 4350
5995 D1 = D9
6000  RETURN

LIST 6001,10000

6010  IF SP = 100 THEN 6070
6015 SP(X) = 0
6020 BB$ = "F": GOTO 5252
6075 SZ(Y) = 0
6080 BB$ = "F": GOTO 5252
6100  FOR FF = 1 TO NS(F)
6110 W(FF) = W
6120  NEXT FF
6130  GOSUB 2035
6140  GOTO 838
6200  PR# 1: PRINT  CHR$ (15)
6300  IF C$ = "114" THEN  PRINT "ALLOWABLE BOND STRESS IS ";C;"N/MM^2"
6330  IF C$ = "110" THEN  PRINT "ULTIMATE BOND STRESS IS ";FB;"N/MM^2"
6350  PRINT "*************************************************************
     *************************************************************"
```

```
6410  PRINT  SPC( 5)"SLAB TYPE"; SPC( 13)"DEPTH"; SPC( 5)"BENDING MOMENT";
      SPC( 5)"LOCAL BOND"; SPC( 5)"BAR SIZE"; SPC( 5)"SPACING"; SPC( 10)"P
      OSITION"
6440  PRINT  SPC( 28)"(MM)"; SPC( 9)"(KN-M)"; SPC( 10)"(N/MM^2)"; SPC( 8)"
      (MM)"; SPC( 9)"MM"
6600  RETURN
6900  PR# 1: PRINT  CHR$ (15)
6910 WRITE = 49568
6930  CALL WRITE:SS$," ",D1 + DS;I3,"          ",MM;F08.3,"          ",QB;F
      5.3,"          ",SZ;I3,"          ",SP;I3,"          ",B$, CHR$ (15):
7100  GOTO 5490
```

14.4 Program I-BASE

I-BASE is a program used to design isolated square bases which support square columns, by the modular-ratio method. The program determines the suitable base size and the amount of reinforcement (including the number and sizes of bars). I-BASE also has a graphics component which can be used to sketch a section through the base showing the number and positions of the reinforcing bars.

14.4.1 Flow chart

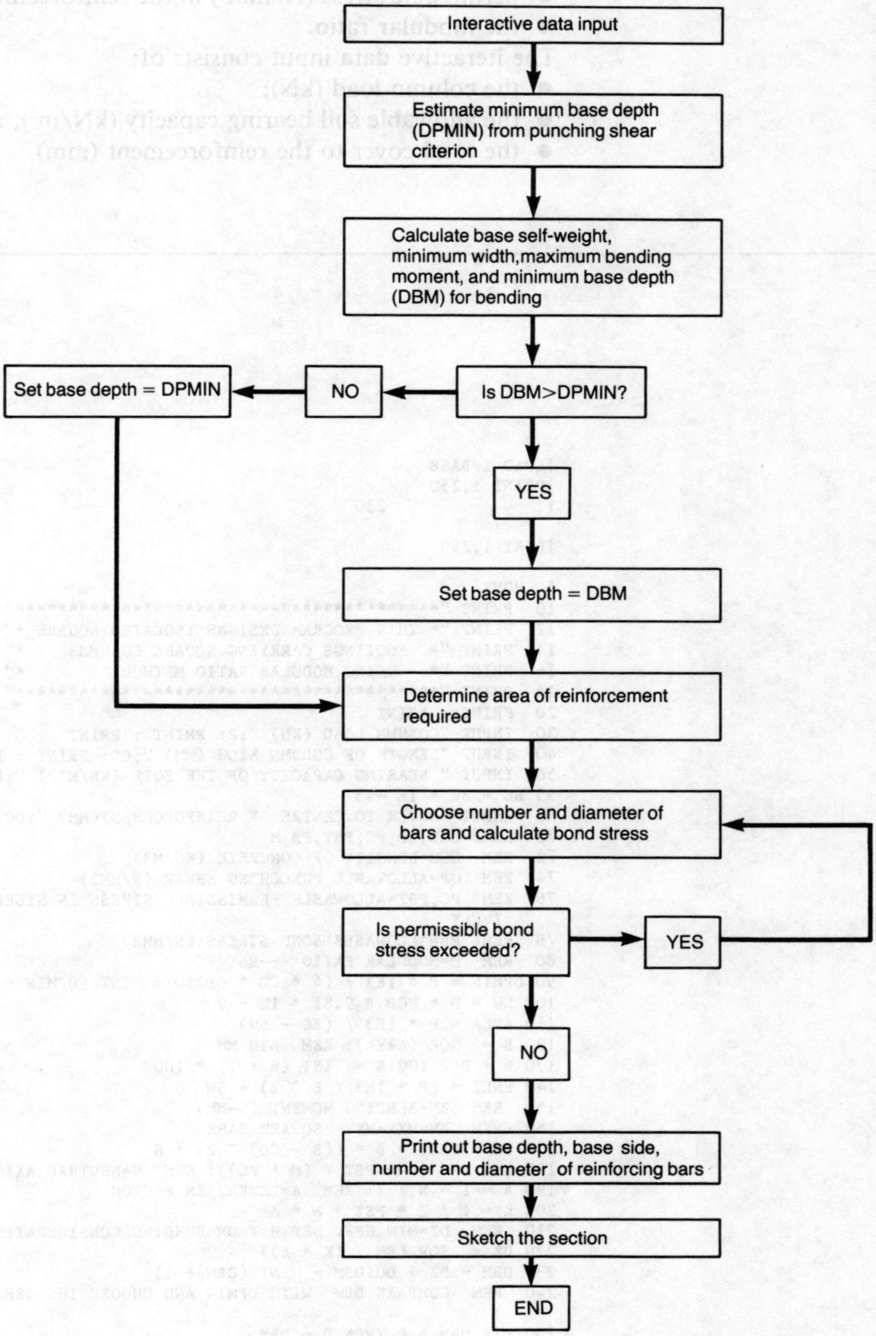

14.4.2 Data input

There are two types of data input for I-BASE, namely:
(i) the **basic** (entered once-and-for-all before program is run); and
(ii) the **interactive** (entered by the user during execution of the program).

The basic data input includes:

- the density (kg/m^2) of concrete;
- permissible stress (N/mm^2) (bending, punching and bond) in the concrete;
- permissible stress (N/mm^2) in the reinforcement; and
- the modular ratio.

The iteractive data input consists of:

- the column load (kN);
- the allowable soil bearing capacity (kN/m^2); and
- the total cover to the reinforcement (mm).

```
]LOAD I-BASE
]PRINT 1,250
1              250

]LIST 1,250

5  HOME
10  PRINT "*****************************************"
12  PRINT "* THIS PROGRAM DESIGNS ISOLATED SQUARE *"
14  PRINT "*  FOOTINGS CARRYING SQUARE COLUMNS    *"
16  PRINT "*   USING MODULAR RATIO METHOD         *"
18  PRINT "*****************************************"
20  PRINT : PRINT
30  INPUT "COLUMN LOAD (KN) ";P: PRINT : PRINT
40  INPUT "LENGTH OF COLUMN SIDE (MM) ";CD: PRINT : PRINT
50  INPUT " BEARING CAPACITY OF THE SOIL (KN/M^2) ";BC: PRINT : PRINT
55 BC = BC * 1E - 3
60  INPUT "COVER TO CENTRE OF REINFORCEMENT(MM) ";DC: PRINT : PRINT
70  READ CCD,QP,PC,PST,PB,M
72  REM  CCD=DENSITY OF CONCRETE (KG/M3)
74  REM  QP=ALLOWABLE PUNCCHING SHEAR (N/MM2)
76  REM  PC,PST=ALLOWABLE PERMISSIBLE STRESS IN STEEL AND CONCRETE RESPECT
    IVELY
78  REM  PB=ALLOWABLE BOND STRESS (N/MM2)
80  REM  M=MODULAR RATIO ---EST/EC
90 DPMIN = P * 1E3 / (4 * CD * QP):D =  INT (DPMIN + 1)
100 SW = D * CCD * 9.81 * 1E - 9
110 AREA = P * 1E3 / (BC - SW)
120 B =  SQR (AREA): REM  BIN MM
130 B = B / 100:B =  INT (B + 1) * 100
140 PNET = (P * 1E3 / B ^ 2) + SW
150  REM  BM=BENDING MOMENT (N-MM)
160  REM  BM=MXX=MYY--SQUARE BASE
170 BM = PNET * B * ((B - CD) ^ 2) / 8
180 N = 1 / (1 + PST / (M + PC)): REM  N=NEUTRAL AXIS FACTOR
190 A = 1 - N / 3: REM  A=LEVER ARM FACTOR
200 K = 1 / 2 * PST * N * A
210  REM  DZ=MIN EFF. DEPTH FROM BENDING CONSIDERATIONS (MM)
220 DZ =  SQR (BM / (K * B))
230 DBM = DZ + DC:DBM =  INT (DBM + 1)
240  REM  COMPARE DBM  WITH DPMIN AND CHOOSE THE GREATER AS BASE DEPTH (MM
    )
250  IF DBM > D THEN D = DBM
```

```
]I-BASE PG 2

?SYNTAX ERROR
]LIST 251,1000

]PR#0
]LIST 251,1000

260 D =  INT (D / 25 + 1) * 25: REM  D TO THE NEAREST 25MM
270 D1 = D - DC
280 AST = BM / (PST * A * D1)
290  PRINT "AST="AST"MM": PRINT : PRINT
300 DIAM = 32: GOSUB 600
310  IF QB < PB THEN 500
320 DIAM = 25: GOSUB 600
330  IF QB < PB THEN 500
340 DIAM = 20: GOSUB 600
350  IF QB < PB THEN 500
360 DIAM = 16: GOSUB 600
370  IF QB < PB THEN 500
380 DIAM = 12: GOSUB 600
390  IF QB < PB THEN 500
500  PRINT "BASE DEPTH="D"MM"; SPC( 6)"BASE SIDE="B"MM"
510  PRINT "USE ";N;"--";DIAM;"MM BARS BOTH WAYS (BTM BARS ONLY)"
515  INPUT "PRESS RETURN TO SKETCH SECTION";R$: GOSUB 670
516  PRINT : PRINT : PRINT
517  INPUT "PRESS RETURN TO CONTINUE";R$
518  TEXT
520  HOME : VTAB (4): INPUT "DO YOU WANT TO DESIGN ANOTHER BASE?(Y/N)";R$:
      IF R$ = "Y" THEN  RESTORE : GOTO 30
530  END : REM  END OF PROGRAM
540  REM  ENTER DATA IN FFLG. SEQUENCE--CCD(KG/M3),QP,PC,PST,PB(N/MM2),M
550  DATA  2440,1.4,7,126,1.25,15
600  REM  THIS SUBROUTINE OBTAINS THE REQD. NO OF BARS AND OBTAINS THE COR
     RESP. BOND STRESS
610 PI = 3.14159
620 N = 4 * AST / (PI * DIAM ^ 2):N =  INT (N + 1)
630 PERIM = N * PI * DIAM
640 Q = PNET * B * (B - CD)
650 QB = Q / (PERIM * A * D1)
660  RETURN
670  REM  THIS SUBROUTINE SKETCHES BASE SECTION
680 D =  INT (D * 200 / B):D1 =  INT (D1 * 200 / B):B = 200
690  HGR
700  COLOR= 1
710  HPLOT 0,0 TO B,0 TO B,D TO 0,D TO 0,0
720 S = B / (N + 1):N = N + 1
730  FOR I = S TO (S * N) STEP S
740  HPLOT I,D1
750  NEXT I
760  RETURN
```

14.5 Program RW-M1

RW-M1 is a special program which can be used to design reinforced cantilever retaining walls. The design can be based on either the modular-ratio or the load-factor method, and the user can choose one of three nominal concrete mixes, ie, 1:1:2; 1:1.5:3; or 1:2:4. The input data consists of:

- the permissible stresses (N/mm^2) (bending, direct shear and local bond) in the concrete; and
- the permissible stress in the reinforcement.

RW-M1 performs automatic checks for sliding, and proceeds to design a rib if necessary. The output consists of the stem thickness for each metre of wall height, and the appropriate reinforcement size and spacing.

The basic assumptions in the program are that:

- the surface of the retained material is horizontal and is flush with the top of the wall;

- there are no externally-applied loads (or surcharge) near the top of the wall that can generate an overturning moment (apart from that due to the retained material);
- there is no load on top of the wall itself that will increase the pressure on the soil below the base;
- the soil is cohesionless;
- the base of the retaining wall has a constant thickness; and
- the water content of the retained material is zero.

14.5.1 Flow chart

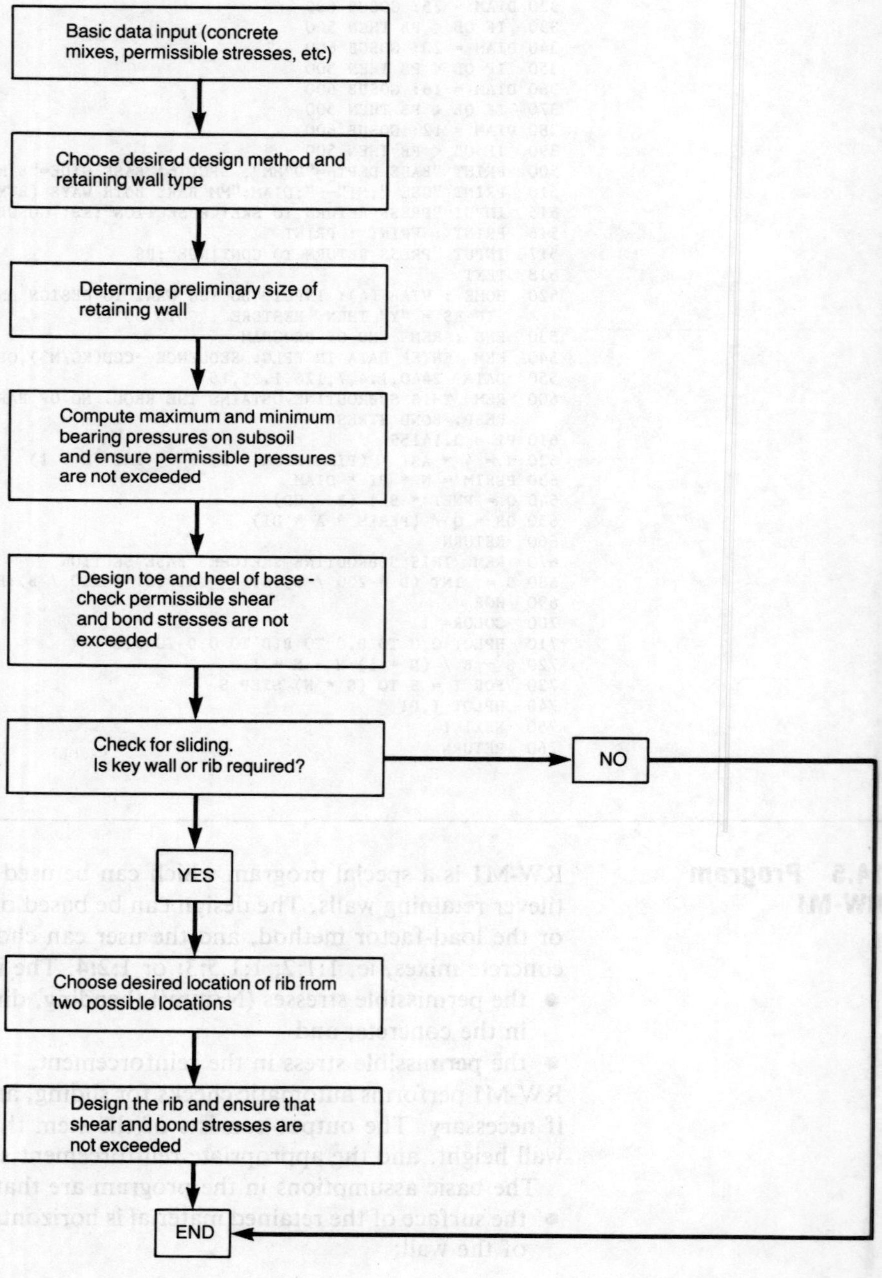

14.5.2 Specimen problem

Design data

Angle of repose	= 30°
Allowable bearing pressure on soil	= 161.0 kN/m^2
Height of material to be retained	= 5.11 m
Unit weight of material to be retained	= 15.71 kN/m^3
Angle of friction of earth on back of the wall	= 0°
Coefficient of friction between base of wall and soil	= 0.40
Depth of earth in front of wall from the point of contact of each with base	= 0.61 m

The top surface of the ground behind the wall is horizontal.

```
]LOAD RW
]LIST 1,170

2  REM      PROGRAM LISTING
15  REM  $$ DESIGN OF REINFORCED CONCRETE                    $$
17  REM  $$ CANTILEVER WALLS                       $$
19  REM  $$ -A MICRO COMPUTER APPROACH             $$
20  REM  SINCE A CONSTANT  THICK
21  REM  $$$$$$$$$$$$$$$$$$$$$$$$$$$$$$$$$$$$$$$$$$$$$$
58  REM  THIS PROGRAM USES THE ´CP 110´ OR THE ´CP 114´ CODE OF PRACTICE
60  REM  LINES 61 TO 65 OUTLINES THE ASSUMPTIONS OF THE PROGRAM
61  REM --THE SOIL IS COHESIONLESS
62  REM --THE WATER LEVEL OF THE SOIL IS ZERO
63  REM --THERE IS NO EXTERNALLY APPLIED LOAD(OR SURCHARGE), EITHER ON THE
      RETAINED MATERIAL (SOIL) OR THE WALL ITSELF
64  REM --THE THICKNESS OF THE BASE IS CONSTANT
65  REM --THE SURFACE OF THE SOIL IS HORIZONTAL AND LEVELS WITH THE TOP OF
      THE WALL
80  REM  LINES 82 TO 178 SHOWS WHAT EACH SYMBOL USED IN THE DESIGNN REPRES
     ENTS (IN ALPHABETICAL ORDER).OTHER SYMBOLS USED IN THE DESIGN ARE DEF
     INED AS THE ANNALYSIS GOES ON
82  REM  ADS=AREA OF DISTRIBUTION BARS
85  REM  BS=BOND STRESS
87  REM  CC=CONCRETE COVER TO REINFORCEMENTS
90  REM  D=EFFETIVE DEPTH
92  REM  DN=DEPTH OF NEUTRAL AXIS
93  REM  DS=DIAMETER OF  DISTRIBUTION BARS
95  REM  E=ECCENTRICITY
97  REM  ER=RESISTANCE PROVIDED BY THE PASSIVE EARTH PRESSURE ACTING AGAIN
     ST THE HEEL
100  REM  F=LEVER ARM FACTOR
102  REM  M,M(H)=BENDING MOMENTS
105  REM  MR=MOMENT OF RESISTANCE
107  REM  MU=MODULAR RATIO
109  REM  LA=LEVER ARM (CP 114) OR LEVER ARM FACTOR (CP 110)
110  REM  N=NUMBER OF MAIN REINFORCING BARS
112  REM  M=BENDING MOMENT (CP 114) OR ULTIMATE BENDING MOMENT (CP 110)
115  REM  O=SUM OF CIRCUMFERENCES
117  REM  OD,OD(H)=OVERALL DEPTH
```

```
120  REM  OD(0)=THICKNESS OF THE STEM AT THE TOP
122  REM  OD(HR)=THICKNESS OF THE STEM AT IT´S JUNCTION WITH THE BASE
125  REM  OH=OVERALL HEIGHT OF THE RETAINING WALL
127  REM  PB=PRESSURE BELOW THE BASE
130  REM  PH=MINIMUM BEARING PRESSURE
132  REM  PP=TOTAL PASSIVE RESISTANCE PER METRE RUN OF THE WALL
135  REM  PR=PRESSURE BELOW THE RIB
137  REM  PT=MAXIMUM BEARING PRESSURE
140  REM  Q=MAXIMUM SHEAR FORCE
142  REM  R=FRICTIONAL RESISTING FORCE
145  REM  RC=UNIT WEIGHT OF REINFORCED CONCRETE
147  REM  RCP=RATE OF CHANGE OF PRESSURE PER LENGTH OF BASE
150  REM  SD=SPACING OF DISTRIBUTION BARS
152  REM  SF=SLIDING FORCE
153  REM  SH=STRESS AT THE JUNCTION BETWEEN THE BASE AND THE TOE
157  REM  SS=SHEAR STRESS
160  REM  T=LENGTH OF TOE
162  REM  TH=DEPTH OF RETAINED MATERIAL IN-FRONT OF WALL + THICKNESS OF BA
     SE + LENGTH OF RIB
165  REM  TW=TOTAL DOWNWARD FORCE
167  REM  WB=WEIGHT OF BASE
170  REM WE=WEIGHT OF RETAINED MATERIAL

LIST 171,320

172  REM  WS=WEIGHT OF STEM
178  REM  Z=LEVER ARM
185  PRINT  CHR$ (15)
190 WRITE = 49568
200  PRINT "FEED IN THE CODE OF PRACTICE (CP) THAT YOU ARE INTERESTED IN (
     114/110)--";
201  INPUT CP: PRINT
202  IF CP = 110 THEN  GOTO 206
204  IF CP = 114 THEN  GOTO 227
206  REM  LINES 210 TO 216 SHOW CONCRETE GRADES (FCU) WITH THEIR CORRESPON
     DING VALUES FOR ´MAXIMUM ALLOWABLE SHEAR STRESS (PQS)´ AND ´ULTIMATE
     LOCAL BONND STRESS (PLBS)´--FOR DEFORMED BARS TYPE 2
208  INPUT "FEED IN THE CONCRETE GRADE (FCU) THAT YOU ARE INTERESTED IN (2
     0/25/30/40/ANY GRADE GREATER THAN 40)--";FCU
210  IF FCU = 20 THEN PQS = 3.35:PLBS = 2.50
212  IF FCU = 25 THEN PQS = 3.75:PLBS = 3.00
214  IF FCU = 30 THEN PQS = 4.10:PLBS = 3.40
215  IF FCU = 40 THEN PQS = 4.75:PLBS = 4.10
222  INPUT "WHAT IS THE CHARACTERISTIC STRENGTH OF THE TENSILE REINFORCEME
     NT-HIGH YIELD STEEL,(N/MM^2) ?--";FY
225  GOTO 265
226  PRINT
227  PRINT "THE NOMINAL CONCRETE MIXES(CEMENT:SAND:GRAVEL) IS CODED AS FOL
     LOWS
228  PRINT "1:1:2 IS REPRESENTED BY THE FIGURE 10 WHILE 1:1.5:3 BY FIGURE
     15"
229  PRINT "ALSO CONCRETE MIX 1:2:4 IS REPRESENTED BY FIGURE 20"
230  PRINT "ANY OTHER CONCRETE MIX APART FROM THOSE MENTIONED ABOVE IS REP
     RESENTED BY 25"
232  PRINT
235  INPUT "FEED IN THE MIX PROPORTIONS THAT YOU ARE INTERESTED IN--";MP
237  PRINT
240  IF MP = 10 THEN PC = 7.6:PQS = 0.9:PLBS = 1.5: GOTO 265
245  IF M = 15 THEN PC = 6.5:PQS = 0.8:PLBS = 1.4: GOTO 265
250  IF MP = 20 THEN PC = 5.3:PQS = 0.7:PLBS = 1 ILLEGAL QUANTITY 25: GOTO
     265
252  IF MP = 25 THEN  GOTO 254
254  INPUT "WHAT IS THE PERMISSIBLE STRESS IN CONCRETE FOR YOUR CHOSEN CON
     CRETE MIX (N/MM^2)?--";PC
255  PRINT
256  INPUT "WHAT IS THE PERMISSIBLE SHEAR STRESS FOR YOUR CHOSEN CONCRETE
     MIX (N/MM^2)?--";PQS
257  PRINT
258  INPUT "WHAT IS THE PERMISSIBLE LOCAL BOND STRESS FOR YOUR CHOSEN CONC
     RETE MIX (N/MM^2)?--";PLBS
265  PRINT
270  INPUT "WHAT IS THE UNIT WEIGHT OF THE RETAINED MATERIAL (N/M^3)?--";W

275  PRINT
```

```
280  INPUT "WHAT IS THE ANGLE OF REPOSE OF THE RETAINED MATERIAL (DEG)?--"
     ;AR
282  PRINT
285  INPUT "WHAT IS THE HEIGHT OF THE RETAINED MATERIAL (M) ?--";HR
290  PRINT
295  INPUT "WHAT IS THE ALLOWABLE BEARING PRESSURE ON THE SUBSOIL (N/MM^2)
     ?--";ABP
300  PRINT
302  INPUT "WHAT IS THE COEFFICIENT OF FRICTION BETWEEN BASE OF WALL AND S
     OIL ?--";CF
304  PRINT


 LIST 306,400

306  INPUT "WHAT IS THE DEPTH OF EARTH IN FRONT OF WALL,FROM THE POINT OF
     CONTACT OF EARTH WITH BASE (M)?--";HE
308  PRINT
310  IF CP = 110 THEN  GOTO 314
312  INPUT "WHAT IS THE PERMISSIBLE STRESS IN HIGH TENSILE STEEL REINFORCE
     MENT (N/MM^2)?--";PST
313  PRINT
314  REM  ASSUMING ABAR DIAMETER,BD OF 25MM
315  REM  THIS PROGRAM ONLY CONSIDERS A UNIT WIDTH ´B´ OF ANY SECTION UNDE
     R CONSIDERATION (B=1M)
316  REM  ´Y=1.4´ IN LINE 317 & 318 IS USED TO FACTOR THE MOMENTS TO ULTIM
     ATE LIMIT STATE
317  READ BD,B,YMU,RC
318  DATA  25,1000,1.4,15,24000
319 BD = 25:B = 1000:MU = 15
320  REM  LINE 325 HOLDS FOR A COHESIONLESS SOIL
322 KA = (1 -  SIN (AR * 22 / (7 * 180))) / (1 +  SIN (AR * 22 / (7 * 180)
     ))
323  PRINT "KA=";KA
324  IF CP = 110 THEN  GOTO 332
325  PRINT "DO YOU WANT TO USE THE LOAD FACTOR METHOD OF DESIGN ? THEN USE
      CODE ´LFM´"
326  PRINT "WHILE IF YOU WANT THE ELASTIC THEORY METHOD DESIGN THEN USE CO
     DE ´ETM´"
327  INPUT "FEED IN THE METHOD OF DESIGN  (CP 114) TTHAT YOU ARE INTERESTE
     D IN--";A$
328  IF A$ = "LFM" THEN K = PC / 4: GOTO 330
329 NA = 1 / (0 + (PST) / (MU * PC)):F = 1 - NA / 3:K = 1 / 2 * PC * NA *
     F
330 K =  INT (K * 1000 + 0.5) / 1000
331  PRINT "K=";K: PRINT
332  PRINT  TAB( 30);"DESIGN FOR STEM OF RETAINING WALL"
333  PRINT  TAB( 30);"**************************************": PRINT
334  PRINT "´H´=> DISTANCE FROM THE TOP OF THE WALL(TABLE BELOW)": PRINT
335  DIM M(HR)
336  PRINT  SPC( 8);"BENDING"; SPC( 4);"EFFECTIVE"; SPC( 4);"OVERALL"; SPC(
     11);"MAIN"; SPC( 18);"DISTRIBUTION"
337  PRINT  SPC( 8);"MOMENT"; SPC( 7);"DEPTH"; SPC( 6);"DEPTH"; SPC( 12);"
     STEEL"; SPC( 19);"STEEL"
338  PRINT  SPC( 11);"M"; SPC( 10);"D"; SPC( 10);"OD"; SPC( 9);"AST"; SPC(
     6);"BD"; SPC( 5);"S"; SPC( 8);"ADS"; SPC( 5);"DS"; SPC( 5);"SD"
339  PRINT  SPC( 8);"(N-M)"; SPC( 8);"(MM)"; SPC( 7);"(MM)"; SPC( 7);"(MM^
     2)"; SPC( 3);"(MM)"; SPC( 3);"(MM)"; SPC( 5);"(MM^2)"; SPC( 2);"(MM)"
     ; SPC( 3);"(MM)"
340  FOR H = 1 TO HR
342 M(H) = W / 6 * H ^ 3 * KA * 1000
345  IF CP = 110 THEN M(H) = Y * M(H)
346  LET M = M(H)
348  LET CC = 200 - 15 * H
349  IF CC < 100 THEN CC = 100
351  IF CP = 110 THEN  GOSUB 1650
353  IF CP = 114 THEN  GOSUB 1420
355  PRINT  SPC( 2);H; SPC( 1);
357  CALL WRITE:M / 1000;F9.1:
359  PRINT  SPC( 4);
361  CALL WRITE:D;F4.1:
363  PRINT  SPC( 6);( INT (OD * 10 + 0.5) / 10); SPC( 6);
365  CALL WRITE:AST;F5.1:
367  PRINT ; SPC( 4);BD; SPC( 4);( INT (S + 0.5)); SPC( 6);ADS; SPC( 4);DS
     ; SPC( 4);( INT (SD + 0.5))
```

```
369  NEXT H
370  PRINT
371  IF HR =  INT (HR) THEN  GOTO 394

]LIST372,410

372 M(HR) = W / 6 * (HR ^ 3) * KA * 1000:M = M(HR)
374  PRINT  TAB( 3);"AT ´H´=";HR;"M FROM THE END OF THE WALL"
376  PRINT  TAB( 3);"---------------------------------------"
378  IF CP = 110 THEN M = Y * M
380  IF CP = 110 THEN  GOSUB 1650
382  IF CP = 114 THEN  GOSUB 1420
384 OD(HR) = OD
386  GOSUB 1990
387  PRINT
390  PRINT  TAB( 2);"SELECTING DIMENSIONS AND THICKNESSES"
391  PRINT  TAB( 2);"************************************"
400  PRINT  TAB( 9);"OF THE RETAINING WALL"
402  PRINT  TAB( 9);"*********************"
403  PRINT
405  PRINT "WHICH OF THE FOLLOWING TYPES OF REINFORCED  CONCRETE RETAINING
      WALL DOYOU WANT TO DESIGN?"
406  PRINT
408  PRINT "IS IT THE TYPE THAT IS SUITABLE  WHEN EXCAVATION BEHIND THE WA
     LL IS NOT POSSIBLE? THEN FEED IN FIGURE 30"
409  PRINT
410  PRINT "IS IT THE MOST EFFICIENT TYPE?, THEN FEED IN FIGURE 35"


]LIST 411,450

411  PRINT
412  PRINT "WHILE FOR THE TYPE, WHERE EXCAVATION BEHIND THE WALL TO FULL L
     ENGTH IS NOT POSSIBLE ,USE 40"
414  PRINT
419  INPUT "FEED IN THE TYPE OF REINFORCED CONCRETE RETAINING WALL THAT YO
     U ARE INTERESTED IN--";TY
420  REM  THICKNESS OF THE BASE, TB ISNOW ASSUMED TO BE OF THE SAME THICKN
     ESS AS THE THICKNESS OF THE JUNCTION OF THE STEM WITH THE BASE
421 TB = OD(HR) / 1000
422 OH = HR + TB:OH =  INT (OH * 1000 + 0.5) / 1000
423 P = W / 2 * OH ^ 2 * KA
424  IF TY = 30 THEN  GOTO 427
425  IF TY = 35 THEN  GOTO 431
426  IF TY = 40 THEN  GOTO 435
427 L = 2 / 3 * OH
429  GOTO 440
431 L =  SQR (2 * P / W)
433  GOTO 440
435 L =  SQR (3 / 2 * P / W)
440 L =  INT (L * 1000 + 0.5) / 1000
441 T = L / 3:T =  INT (T * 1000 + 0.5) / 1000
443  REM  THE FIGURE ´1000´ IN LINES 445,455 & 460 IS ACONVERSION FACTOR
445  LET OD(0) = (OD(1) - 13) / 1000:OD(0) =  INT (OD(0) * 1000 + 0.5) / 1
     000

LIST 451,918

553  LET LB = L - T - TB
560 WS = 1 / 2 * (OD(0) + TB) * HR * RC
570 WB = L * TB * RC
580 WE = HR * LB * W
590 TW = WS + WE + WB
610  REM  LINE 620 SHOWS THE MOMENT ABOUT THE ENTRE-LINE OF THE BASE,MB
615  PRINT
620 MB = (P * OH) / 3 + WS * (L / 2 - (L / 3 + TB / 2)) + WE * (L / 2 - (L
     B / 2 + T + TB))
640 E = MB / TW
700  IF E < L / 6 THEN  GOTO 750
710  IF E = L / 6 THEN  GOTO 780
720  IF E > L / 6 THEN  GOTO 810
730  REM  DETERMINING THE BEARING PRESSURES ON THE SUBSOIL AND  CHECKING W
     HETHER THE BEARING PRESSURE, PT IS LESS THAN THE ALLOWABLE BEARING PR
     ESSURE
740  REM  EXPRESSIONS ON LINES 750 TO 810 ARE FOR A UNIT WIDTH, (1M)
```

```
750 PT = TW / L + (6 * TW * E) / (L ^ 2)
760 PH = TW / L - (6 * TW * E) / (L ^ 2):PH =  SGN (PH) *  INT ( ABS (PH) *
    10 + 0.5) / 10
770  GOTO 816
780 PT = 2 * TW / L
790 PH = 0
800  GOTO 816
810 PT = (4 * TW) / (3 * E * L)
815 PH = 0
816 PT =  SGN (PT) *  INT ( ABS (PT) * 10 + 0.5) / 10
817  IF PT < ABP THEN  GOTO 824
819  REM  THE LENGTH OF THE BASE,L IS NOW INCREASED  BY 5MM
820  LET L = L + 5 / 1000
824 TB =  INT (TB * 1000 + 0.5) / 1000: PRINT
825  PRINT "THICKNESS OF THE BASE, TB=";TB"M"
826  PRINT
827  PRINT "OVERALL HEIGHT OF THE RETAINING WALL,OH=";OH;"M"
828  PRINT
829  PRINT "LENGTH OF BASE,L=";L;"M"
830  PRINT
831  PRINT "LENGTH OF THE TOE,T=";T;"M"
832  PRINT
833  PRINT "THICKNESS OF THE STEM AT THE TOP,OD(0)=";OD(0);"M"
834  PRINT
836  PRINT "MAXIMUM BEARING PRESSURE,PT=";PT;"N/MM^2"
837  PRINT
840  PRINT "MINIMUM BEARING PRESSURE,PH=";PH;"N/M^2"
841  PRINT
845  PRINT "THE MAXIMUM BEARING PRESSURE IS OKAY"
850 RCP = (PT - PH) / L
860 SH = PT - (T + TB) * RCP
870 ST = PT - (T * RCP)
880  PRINT
885  PRINT
890  PRINT  TAB( 9);"DESIGN FOR THE TOE OF THE BASE"
900  PRINT  TAB( 9);"******************************"
905  PRINT
910 M = (ST * T * T / 2 + (PT - ST) * (T ^ 2) / 3 - T * TB * RC * T / 2) *
     1000
912  IF CP = 110 THEN M = Y * M
916 Q = (ST * T) + T * (PT - ST) / 2 - T * TB * RC
918  IF CP = 110 THEN Q = Y * Q

 LIST 919,1300

920  REM  SINCE ACONSTANT THICKNESS OF BASE WAS ASSUMED,THEREFORE THE THIC
     KNESS OF THE TOE AND THE HEEL IS THE SAME AS THAT OF THE BASE,TB. HEN
     CE
921  REM  FIGURE ´25´ IN LINES 922,980,1540 & 1660 STANDS FOR THE ASSUMED
     DIAMETER OF BARS
922 D = TB * 1000 - CC - 25 / 2:OD = TB * 1000
923  IF CP = 110 THEN  GOSUB 1665
924  IF A$ = "ETM" THEN  GOSUB 1570
925  IF A$ = "LFM" THEN LA = 0.75 * D:AST = M / (PST * LA): GOSUB 1695
926  GOSUB 1990
928  GOSUB 2060
929  PRINT
930  PRINT  TAB( 9);"DESIGN OF THE HEEL OF THE BASE"
940  PRINT  TAB( 9);"******************************"
945  PRINT
950 M = ((WE + LB * TB * RC) * LB / 2 - PH * (LB ^ 2) / 2 - (SH - PH) * (L
     B ^ 2) / 6) * 1000
955  IF CP = 110 THEN M = Y * M
970 Q = (WE + LB * TB * RC) - PH * LB - (SH - PH) * LB / 2
975  IF CP = 110 THEN Q = Y * Q
980 D = TB * 1000 - CC - 25 / 2:OD = TB * 1000
985  IF CP = 110 THEN  GOSUB 1665
990  IF A$ = "ETM" THEN  GOSUB 1570
995  IF A$ = "LFM" THEN LA = 0.75 * D:AST = M / (PST * LA): GOSUB 1695
1000  GOSUB 1990
1010  GOSUB 2060
1040  PRINT  TAB( 11);"CHECK FOR SLIDING"
1050  PRINT  TAB( 11);"*****************"
1055  PRINT
1060  LET FOS = 2
```

```
1070 SF = FOS * P
1080 R = CF * TW
1084 FR = SF - R
1090  IF SF > R THEN  GOTO 1120
1100  PRINT "A KEY WALL OR RIB IS NOT REQUIRED"
1114  PRINT
1115  HTAB (14): FLASH : PRINT "END OF RESULTS"
1116  NORMAL
1118  END
1120  PRINT "A KEY WALL OR RIB IS REQUIRED"
1125  PRINT
1130  PRINT  TAB( 13);"DESIGN OF RIB"
1140  PRINT  TAB( 13);"*************"
1145  PRINT
1160  PRINT "IF YOU WANT THE LOCATION OF THE RIB TO BE AT THE HEEL OF THE
     BASE SLAB,THEN FEED IN CODE ´RH´"
1170  PRINT "WHILE I AT THE TOE OF THE BASE SLAB ,THEN USE ODE ´RT´"
1172  PRINT
1175  INPUT "FEED IN THE LOCATION OF RIB THAT YOU ARE INTERESTED IN--";B$
1180  IF B$ = "RH" THEN  GOTO 1200
1190  IF B$ = "RT" THEN  GOTO 1310
1195  PRINT
1200  REM  FOR KEY WALL (RIB) AT FRONT OF HEEL
1230  LET ER = 2 / 3 * ABP
1240 LR = (FR * 1000) / ER
1260 M = FR * LR / 2
1275  PRINT
1280 Q = FR
1290  GOTO 1372


 LIST 1301,1720

1310  REM  FOR KEY RIB AT FRONT OF TOE
1320 Q = FR:PP = Q
1330 TH =  SQR (2 * PP * KA / W)
1340 LR = (TH - HE) * 1000
1348  PRINT
1350 PB = HE * W / KA
1360 PR = TH * W / KA
1370 M = (PB * LR / 1000 * (LR / 1000) / 2 + (PR - PB) * ((LR / 1000) ^ 2)
     / 3) * 1000
1372  IF LR > 4 * OD(HR) THEN LR = 4 * OD(HR)
1373 LR =  SGN (LR) *  INT (LR * 10 + 0.5) / 10
1375  PRINT "LENGTH OF THE RIB,LR=";LR;"MM"
1380  PRINT
1388  IF CP = 110 THEN  GOSUB 1650
1390  IF CP = 114 THEN  GOSUB 1420
1392  IF OD > 3 / 4 * OD(HR) THEN  GOTO 1400
1394 OD = 3 / 4 * OD(HR):D = OD - CC - BD / 2
1396  IF CP = 110 THEN  GOSUB 1665
1398  IF CP = 114 THEN  GOSUB 1570
1400  GOSUB 1990
1402  GOSUB 2060
1404  PRINT
1410  HTAB (14): FLASH : PRINT "END OF RESULTS"
1412  NORMAL
1413  PRINT  CHR$ (18)
1415  END
1420  REM  THE FOLLOWING PROGRAM SHOWS THE SUBROUTINES IN THE DESIGN OF RE
     TAINING WALL
1510 D =  SQR (M / (K * B))
1515  GOTO 1540
1520  REM  THE EFFECTIVE DEPTH,IS NOW INCREASED BY 2MM
1530  LET D = D + 2
1535  REM  FIGURE ´25´ IN LINES 922,980,1540 & 1660
1540 OD = D + CC + 25 / 2
1550  LET OD(H) = OD
1570  REM  LINES 1580 TO 1760 SHOWS THE PROGRAM FOR THE COMPUTATION THE AR
     EA OF REINFORCEMENTS
1580 MR = K * B * D ^ 2
1590  REM  LINE 1600 INDICATES THAT ONLY TENSILE REINFORCEMENT WILL BE PRO
     VIDED
1600  IF MR > M THEN  GOTO 1620
1610  GOTO 1520
```

```
1620  IF A$ = "LFM" THEN AST = (PC * B * D) / (3 * PST):DN = (3 * AST * PS
     T) / (2 * B * PC):LA = D - DN / 2: GOTO 1695
1630 LA = F * D
1640 AST = M / (PST * LA): GOTO 1695
1650 D =  SQR (M / (0.15 * FCU * B))
1660 OD = D + CC + 25 / 2:OD(H) = OD
1665  LET C = M / (B * D ^ 2 * FCU)
1670 LA = 1 / 2 +  SQR (1 / 4 - C / 0.89):Z = LA * D
1675 AST = M / (0.87 * FY * Z)
1680  REM  LINE 1685 SHOWS THE MINIMUM AREA OF TENSILE REINFORCEMENT (MAST
     ) FOR A HIGH YIELD STEEL
1685 MAST = (0.15 * B * D) / 100
1690  IF AST < MAST THEN AST = MAST
1695  IF AST < 500 THEN BD = 12: GOTO 1715
1700  IF AST < 1100 THEN BD = 16: GOTO 1715
1705  IF AST < 2000 THEN BD = 20: GOTO 1715
1710  IF AST > 2000 THEN BD = 25
1715 AST =  INT (AST * 10 + 0.5) / 10
1718 N = AST / (22 / 7 * (BD ^ 2) / 4)
1720 S = 1000 / N
```

```
 LIST 1721,10000

1725  IF S < 100 THEN S = 100
1730  IF S > 300 THEN S = 300
1735 ADS = (0.12 * B * OD) / 100:ADS =  INT (ADS * 10 + 0.5) / 10
1736  IF ADS <  = 500 THEN DS = 10
1737  IF ADS > 500 THEN DS = 12
1740 SD = (1000 * 22 / 7 * (DS ^ 2) / 4) / ADS
1745  IF SD < 100 THEN SD = 100
1750  IF SD > 300 THEN SD = 300
1760  REM  "CHECK FOR SHEAR"
1770  IF CP = 114 THEN  GOTO 1860
1780 SS = Q / (B * D)
1790  IF SS < PQS THEN  GOTO 1910
1800  LET D = D + 1
1810  GOTO 1660
1860 SS = Q / (B * LA)
1880  IF SS < PQS THEN  GOTO 1910
1890  GOTO 1520
1910 SS =  INT (SS * 100 + 0.5) / 100
1915  REM  CHECK FOR BOND
1920 O = 22 / 7 * N * BD
1930  IF CP = 114 THEN BS = Q / (O * LA)
1940  IF CP = 110 THEN BS = Q / (O * D)
1950  IF BS < PLBS THEN  GOTO 1982
1960  LET N = N + 1
1970  GOTO 1720
1982 BS =  INT (BS * 100 + 0.5) / 100
1985  RETURN
1990  PRINT "BENDING MOMENT=";M;"N-MM"
1991  PRINT
1993  PRINT "EFFECTIVE DEPTH,D=";
1994  CALL WRITE:D;F4.1:
1995  PRINT ;: PRINT "MM"
1996  PRINT
2000  PRINT "OVERALL DEPTH,OD=";
2002  CALL WRITE:OD;F4.1:
2004  PRINT ;: PRINT "MM"
2010  PRINT "AREA OF MAIN REINFORCEMENT,AST=";AST;"MM^2"
2015  PRINT
2020  PRINT "PROVIDE T";BD;"BARS AT";
2021  CALL WRITE:S;I3:
2022  PRINT ;: PRINT "MM CENTRES"
2025  PRINT
2030  PRINT "AREA OF DISTRIBUTION STEEL,ADS=";ADS;"MM^2"
2035  PRINT
2040  PRINT "PROVIDE T";DS;" BARS AT";
2041  CALL WRITE:SD;I3:
2043  PRINT ;: PRINT "MM CENTRES"
2055  RETURN
2058  PRINT
2060  PRINT "SHEAR STRESS,SS=";SS;"N/MM^2"
2065  PRINT
```

```
2070  PRINT "THE SECTION IS OKAY FOR SHEAR"
2080  PRINT
2090  PRINT "BOND STRESS, BS=";BS;"N/MM^2"
2100  PRINT
2110  PRINT "THE SECTION IS OKAY FOR  BOND"
2120  PRINT
2130  RETURN

]PR#1
```

References

1. *CP 114: The Structural Use of Reinforced Concrete in Buildings.* British Standards Institute, London, 1969.
2. *CP 110: Code of Practice for the Structural Use of Concrete.* British Standards Institute, London, 1972.
3. Desai, C.S. and Abel, J.F. *Introduction to the Finite Element Method.* Van Nostrand Reinhold Co, New York, 1972.
4. Gallagher, R.H. *Finite Element Analysis.* Prentice-Hall Inc, New Jersey, 1978.
5. Lampert, P. and Collins, M.P. "Torsion, bending and confusion — an attempt to establish the facts" *Proceedings, ACI, 69*(8), 1972.
6. *NCP 1 (Part 1): The Structural Use of Concrete in Buildings.* Nigerian Standards Organisation, Nigeria, 1973.
7. Oladapo, I.O. *Fundamentals of the Design of Concrete Structures.* Evans Brothers Ltd, Ibadan, Nigeria, 1981.
8. Orangun, C.O. "Adaptability of Limit State Design to Local Reinforced Concrete Code of Practice". *The Nigerian Engineer*, Vol. 11, 1976.
9. Terzaghi, K. and Peck, R.B. *Soil Mechanics in Engineering Practice.* second edition, John Wiley, New York, 1967.
10. Zienkiewicz, O.C. *The Finite Element Method.* McGraw-Hill (UK) Ltd, 1977.

Bibliography

Allen, A.H. *Reinforced Concrete Design to CP 110 — simply explained.* Cement and Concrete Association, London, 1974.

Boughton, B. *Reinforced Concrete Detailer's Manual.* Crosby Lockwood Staples, London, 1969.

Everard, N.J. and Tanner III, J.L. *Reinforced Concrete Design.* McGraw-Hill, New York, 1966.

Mosley, W.H. and Bungey, J.H. *Reinforced Concrete Design*, Macmillan, London, 1976.

Reynolds, C.E. and Steedman, J.C. *Reinforced Concrete Designer's Handbook*, A Viewpoint Publication, 1974.

Kong, F.K. and Evans, R.H. *Reinforced and Prestressed Concrete*, Van Nostrand Reinhold (UK) Co. Ltd, 1975.

Model Answers

The following are model answers to some (about one quarter) of the exercises in the textbook. As these relate basically to design exercises, the answers should be viewed as essentially indicative of the range of results to be expected. Where appropriate comments on how the answers have been obtained are provided.

Chapter 1

1.1	d	**1.11**	b	**1.21**	b
1.2	c	**1.12**	c	**1.22**	b
1.3	c	**1.13**	c	**1.23**	c
1.4	a	**1.14**	d	**1.24**	b
1.5	c	**1.15**	a	**1.25**	d
1.6	d	**1.16**	a	**1.26**	c
1.7	d	**1.17**	b	**1.27**	b
1.8	d	**1.18**	b	**1.28**	d
1.9	d	**1.19**	a	**1.29**	b
1.10	d	**1.20**	d	**1.30**	d

Chapter 3

3.1 Using 40 mm total cover:
overall depth = 510 mm (solution obtained by iterations)
width = 255 mm
number of 20 mm ϕ bars required = 5

3.4 (a) $A_{st} = \dfrac{(D_0^3 - D_1^3)}{3m(D_0 + D_1)}$

(b) $M_R = \dfrac{p_c}{8D_0}\left[\dfrac{\pi(D_0^4 - D_1^4)}{8} + \dfrac{(D_0 + D_1)(D_0^3 - D_1^3)}{3}\right]$

(c) The lever arm for the hollow circular beam is (after simplifying):

$$z = \frac{3(D_0 + D_1)}{4}\left[\frac{\pi(D_0^2 + D_1^2)}{8(D_0^2 + D_0 D_1 + D_1^2)} + \frac{1}{3}\right]$$

Therefore, taking limits as $D_0, D_1 \to D = \frac{1}{2}(D_0 + D_1)$, gives:

$$Z = \frac{1}{2}(\pi/4 + 1)D \simeq 0.8927D$$

3.9 (a) $A_{st} = M/p_{st}z$

where:

$$z = \frac{\frac{1}{4}\beta_0 x^2 + \frac{1}{3}[\beta_0(d-x) + \beta_1]x + \frac{1}{2}\beta_1(d-x)}{\frac{1}{3}\beta_0 x + \frac{1}{2}\beta_1}$$

$$\beta_0 = (1-\alpha)b/d$$

$$\beta_1 = (1-\alpha)b(1-x/d) + \alpha b$$

and

$$x = d/(1 + p_{st}/mp_c)$$

(b) $\beta = z'/z$

where

$$z' = d - \frac{1}{3}x$$

(c) Seven 32 mm ϕ bars are required.

3.12 Percentage error $= \dfrac{100b_w(x-h_f)^2(3d-2h_f-x)}{3bx^2(d-x/3) - 6(b-b_w)(x-h_f)^2(d - \frac{2}{3}x - h_f)}$

where x is as in equation (3.43).

The required condition is for:

$$b_w'(n-h_f')(3-2h_f' - n) < 0.06[\tfrac{1}{2}b'n^2(1-n/3) - (b'-b_w')(n-h_f')^2(1-2n/3 - h_f')]$$

where $b_w' = b_w/d$, $h_f' = h_f/d$, $b' = b/d$ and $n = x/d$.

Chapter 4

4.1 Assumed total load $= 28\ \text{kN/m}^2$

(a) Provide a 300 mm thick slab (this is based principally on the desire to reduce the amount of reinforcement).
Use 12 mm ϕ bars at 125 mm c/c (bottom) along the shorter span and for distribution.
Use 8 mm ϕ bars at 200 mm c/c.

(b) Adopting the same 300 mm slab as in (a):
use 12 mm ϕ bars at 200 mm c/c (bottom) along the shorter span and 10 mm ϕ bars at 200 mm c/c (bottom) along the longer span.

4.6 Assumed total dead load $= 4.93\ \text{kN/m}^2$
Assumed total imposed load $= 3.00\ \text{kN/m}^2$

Provide a 90 mm thick waist.

At supports $\begin{cases} \text{Use 12 mm } \phi \text{ bars at 175 mm c/c} \\ \text{and 6 mm } \phi \text{ bars at 175 mm c/c} \end{cases}$

Within span $\begin{cases} \text{Use 12 mm } \phi \text{ bars at 200 mm c/c} \\ \text{and 6 mm } \phi \text{ bars at 200 mm c/c} \end{cases}$

Chapter 5

5.1 Provide four 20 mm ϕ bars. Use 6 mm ϕ bars at 240 mm pitch for transverse reinforcement.

5.6 10.8 kN/m (based on the condition that no part of the column should be in tension).

Chapter 6

6.1 Provide a 2.6 m × 2.6 m × 510 mm base (punching shear consideration is the most critical).
For reinforcement, use 26 no. 10 mm ϕ bars both ways (bend check is crucial).

6.3 For assumed stresses p_{st} = 126 N/mm, p_c = 7 N/mm^2, provide a 4.4 m × 2.9 m × 480 mm base (punching shear consideration is the most critical).
For reinforcement, use:
26 no. 16 mm ϕ bars (bottom) along the shorter side and
18 no. 16 mm ϕ bars (bottom) along the longer side.

Chapter 7

7.4 $\dfrac{d_L}{d_M} = \sqrt{[2\beta(3-\beta)/3]}$

$$\frac{A_L}{A_M} = \frac{4}{9}\sqrt{[3(3-\beta)/2\beta]}$$

where $\beta = 1/(1+p_{st}/mp_c)$.

Chapter 9

9.1 M_u = 623.0 kN m.

9.6 Use four 32 mm ϕ bars.

9.7 With f_{cu} = 40 N/mm^2 and f_y = 410 N/mm^2, provide a 320 mm deep beam and use five 40 mm ϕ bars.

Chapter 11

11.1 $N = 2550$ kN

11.6 Provide a 800 mm × 600 mm column with four 25 mm ϕ bars.

Chapter 12

12.1 Provide a 560 mm deep beam. Use 68 no. 5 mm ϕ wires.

Index